STUDY GUIDE

DIXIE J. GOSS
Hunter College, New York

GENERAL CHEMISTRY

FOURTH EDITION

HILL • PETRUCCI • McCREARY • PERRY

Upper Saddle River, NJ 07458

Project Manager: Kristen Kaiser
Senior Editor: Kent Porter-Hamann
Editor-in-Chief, Science: John Challice
Vice President of Production & Manufacturing: David W. Riccardi
Executive Managing Editor: Kathleen Schiaparelli
Assistant Managing Editor: Becca Richter
Production Editor: Jeffrey Rydell
Supplement Cover Manager: Paul Gourhan
Supplement Cover Designer: Joanne Alexandris
Manufacturing Buyer: Ilene Kahn
Cover Image Credit: MgO(100) Surface/Royce Copenheaver

Pearson Prentice Hall
Pearson Education, Inc.
Upper Saddle River, NJ 07458

Printed in the United States of America

10 9 8 7 6 5 4

ISBN 0-13-140347-8

Pearson Education Ltd., *London*
Pearson Education Australia Pty. Ltd., *Sydney*
Pearson Education Singapore, Pte. Ltd.
Pearson Education North Asia Ltd., *Hong Kong*
Pearson Education Canada, Inc., *Toronto*
Pearson Educación de Mexico, S.A. de C.V.
Pearson Education—Japan, *Tokyo*
Pearson Education Malaysia, Pte. Ltd.

Table of Contents

Introduction

This study guide is designed to help you understand and appreciate chemistry. Chemistry offers many challenges to the beginning student. Some exercises will require memorization and some will require new ways of thinking about problems and problem solving. Much of chemistry attempts to explain observed interactions and reactions. From these explanations, the next step is to predict other interactions and reactions. This phenomenon is not always linear, often we start in the middle and work our way out towards more detailed explanations and predictions. Someone unfamiliar with basketball may observe the game in much the same "nonlinear" way. The first observation might be that there are two teams of five members each. Trying to figure out the objective of the game (scoring) and how this is accomplished (putting the ball through the basket) would come next. After these general observations, the many subtleties of the game, such as how the ball is moved down the court and the unique moves that allow players to position themselves for shots, might be considered.

A chemist may observe that two substances react. She might be interested in the product of the reaction and then proceed to more details, such as formulating models about the types of atoms and molecules and the changes that they undergo. A chemist is unable to see an atom and cannot count the number of molecules. Similarly you cannot watch an electron move. While a physics student may study blocks or balls rolling down planes or traveling through trajectories, chemistry requires the use of abstract models. A model is a simplified way of picturing a reality. As our knowledge of chemistry has increased over the years, models have changed to accommodate this new information.

Hints for studying chemistry

Don't be overwhelmed. Chemistry moves at a very fast pace and it is important to keep up. Don't wait for understanding to strike before starting the problems. The best way is to start and keep going.

1. Read the material before the lecture. You may also want to look at the study guide to pick out the key points of the material before the lecture.
2. Take notes of the key points and explanations during the lecture, but also listen and try to understand the explanations. When the lecture has explained something you had difficulty understanding before, make notes afterward to remind yourself of the concept.
3. Do the problems. Attempt some of the easier problems before the lecture. This will help you to determine where you need more work and what to pay careful attention to. It cannot be overemphasized that chemistry is learned by doing. You will not understand concepts and applications by reading the solutions manual or watching your instructor solve problems or by watching your fellow students solve problems. YOU must work problems yourself.
4. When working problems try to think through the problem first before beginning work. What is being asked in the problem? What information is given? Often by thinking through a problem before simply plugging in lots of numbers you can save time and gain a better understanding of the material.

5. Write out solutions to the problems in a clear fashion just as you would present them in answers to an exam. If you are using a calculator for exams, use it for problems and become thoroughly familiar with its operation.
6. Don't spend too much time on any one problem. If you haven't gotten a good start in 15 minutes or so move on to another problem, then get help from the instructor or another student, or check the solutions manual if it is available. If you do use the solutions manual be sure you do another similar problem without looking up the answer. Reading the answers doesn't work either! You have to do it yourself.
7. Find a study group or study partner. Often explaining the answer to someone else clarifies any points you are uncertain of and helps to improve your understanding. However, don't let someone else do all the work.

Using Your Calculator

You need to be familiar with your particular calculator before the first exam. The best way to do this is to use your calculator to solve problems and in the lab. First take time to look at the instruction book for your calculator. Usually a number of examples are given and the features are explained. Try some of the examples below to be sure you understand how to use your calculator.

Entering Data

Data is usually entered into calculators in one of two different methods: The first method resembles computer programming (CP) and the second uses algebraic notation. CP calculators use an ENTER key. To add 3 and 7 using a CP calculator the sequence is : 3 ENTER 7 +; to multiply the sequence is 3 ENTER 7 x. On a calculator with algebraic notation the equation is entered directly into the calculator. Calculators using algebraic notation use an equals (=) key. The key sequence to add 3 and 7 is: 3 + 7 = and multiplication is 3 x 7 =.

Operations

You can carry out a sequence of operations on your calculator, but each calculator has its own priorities for operations. If you have an algebraic calculator try this key sequence: 2 + 4 x 5 = . Did you get 22? If so, your calculator completes multiplication (and division) before addition (or subtraction). That is, 2 + (4 x 5) = 22. You can change the priority of operations by using parenthesis. The calculator will perform the operation inside the parentheses first. For example (2 + 4) x 5 = 30. Some calculators will carry out the operations in the order listed. For example 2 + 4 x 5 = will give 30. In this case you need to use parentheses around the multiplication if that is the first operation: 2 + (4 x 5) =

You need to use care when doing multiple operations. When stringing together multiplication and division for example:

(20 x 2.00 x 4.00)/(1.5 x 8.00 x 10) =

In this case you can multiply 20 x 2 x 4.00 and divide successively by 1.5, 8.00 and 10. With these types of operations, as with other problems, it is particularly useful to estimate the answer. For example, the numerator is 160, divide by 10 = 16. Divide by 8 gives 2.

Divide by 1.5 gives 4/3. For other examples, approximate the numbers to get something you can estimate. For example:

(1.9732 x 7.741)/(4.036 x 16.32) This can be approximated as (2 x 8)/(4 x 16) which is 1/2 x 1/2 or approximately 1/4 or 0.25. The actual answer is 0.2319. If you got 61.76, what did you do wrong?

Other operations you will need

Learn to take logarithms to the base 10: LOG and base e: LN.

Raise 10 to a power: 10^x or INV LOG

Raise a number to a power: 5^2: 5 y^x 2 =

Invert a number: 1/x

Take squares: x^2 and square roots: $\sqrt{x}$.

Learn to enter scientific notation: 3.1×10^5 Use 3.1 EXP 5. Do not use the x key. Your calculator reads 3.1 05. To enter 3.1×10^{-5} use the key sequence 3.1 EXP +/- (or CHS) 5. Your calculator reads 3.1 -05. If you use the +/- or CHS key before the number, you change the sign of the number. If you use it before the exponent, you change only the exponent.

CHAPTER 1

Chemistry: Matter and Measurement

Chapter Objectives:

In this chapter you will learn some of the key terms, the "language of chemistry", a standard system of scientific measurements, and problem solving methods. These tools are necessary to describe chemical systems and solve scientific problems.

You should be able to:

1. Define and use the terms listed in "Key Terms".
2. Describe the difference between chemical and physical properties and chemical and physical change.
3. Classify matter.
4. Define scientific hypothesis and scientific law.
5. Know the common units of the metric system for volume, length, and mass, and the common prefixes.
6. Define temperature and convert between temperature in Celsius, Kelvin and Fahrenheit.
7. Know the difference between precise and accurate measurements and how to express the degree of precision with the correct number of significant figures.
8. Understand the unit conversion method of problem solving and convert between English and metric units.
9. Estimate the answers to problems.

Chapter Summary

1. Defining and using the terms listed in "Key Terms."

This is a memorization task. You may wish to use 3 x 5 index cards with the term on one side and the definition on the other, or a tape recorder may be useful. To use a tape recorder, record the word and allow time for you to say the definition, or use the pause button. Then use the word in a sentence. You will find definitions in the back of the book, but write the definition in your own words. The definition should be something you can understand and use. Consider the term "data." **Data** could be defined as measurements made during an experiment. What does this mean? If you want to collect data on the boiling point of several liquids, you measure the temperature on a thermometer at which boiling occurs. These numbers are the data. Now try using the term in a sentence. We weighed the samples in our chemistry lab. These data (the weights we measured) were used to determine the amount of material that reacted. (Data is plural, datum is singular.)

2. Differences between chemical and physical properties and chemical and physical changes.

Think of everyday events and objects. Which are **chemical properties** (reacting an item with another produces a new substance) and which are **physical properties** (causes no change in composition). For example, cookie dough is brown (physical), soft (physical), can be divided into a number of pieces (physical), bakes into cookies (chemical), burns (chemical), and can be digested to produce energy (chemical). Think of some other examples, such as mowing the grass (the grass clippings forming compost, mixing the compost in the garden, etc.).

Example 1.1. Which of the following are chemical changes and which are physical changes?

a. Liquid acetic acid freezes to a solid in a cold room.

This is a physical change since the substance is changing state, but no change in chemical composition results. The acid could be thawed and would be the same as before freezing.

b. A sample of magnesium burns brightly in air to form a white powder called magnesium oxide.

Here a chemical change has occurred. The magnesium has combined with oxygen to form a new chemical compound with different properties.

c. A dull knife blade is sharpened with a grindstone.

This is a physical change. The steel of the knife has been physically removed to give a sharp edge, but the steel has not changed chemical properties.

3. Classifying matter. Examples of a substance, a homogeneous mixture, a heterogeneous mixture, a compound and an element.

For each of the descriptions think of your own examples. **Substances** have a definite composition that does not vary from one sample to another. Pure substances are of two types: **elements** that are incapable of being broken down chemically, and **compounds** that contain elements in a fixed proportion. Oxygen is an element while carbon dioxide (dry ice in its solid form) is a compound. Carbon dioxide contains one carbon atom and two oxygen atoms in each molecule. The ratio is always the same. **Mixtures** have varying proportions of substances. A **homogeneous mixture** has the same composition throughout, but may vary from sample to sample. Sugar in a cup of coffee is an example. Assuming the sugar is dissolved, this is a homogeneous mixture. However, if you do not stir the coffee and sugar settles to the bottom so that the composition is not uniform, then this is a heterogeneous mixture. Pick a number of common items and classify them. Select simple items: ink in your pen, an aluminum can, fruit salad, and so on.

Example 1.2. Classify the following according to whether they are substances, homogeneous mixtures or heterogeneous mixtures.

a. Oxygen gas

This is a substance. Oxygen is an element and cannot be separated. However, oxygen gas is O_2, a molecule containing two atoms of oxygen, and is one of the two common allotropes of this element.

b. Orange juice

Orange juice is a heterogeneous mixture. You have to shake the container to get a more uniform distribution of the contents.

c. Fruit salad

This is a heterogeneous mixture. You can pick out the separate components.

d. Cola softdrink

The distribution is usually uniform throughout, so this is a homogeneous mixture.

e. Toothpaste

This is a little tricky and may depend on what brand you use, but food processors would like this to be a uniform composition and it is therefore a homogeneous mixture. If however, you are thinking about one of those toothpastes that has a stripe of gel in it, then it is a heterogeneous mixture.

4. Scientific Method. Define "hypothesis" and "scientific law."

Think of several hypotheses and scientific laws. Plants grow better in sunlight. How can you test this? What other explanations might account for the observation? Perhaps sunny rooms are warmer. Hypotheses cannot be proven, only disproven. If the facts are inconsistent with the hypothesis, we must discard the hypothesis. A hypothesis seeks to explain an observed phenomenon, but a future test may disprove the hypothesis. A hypothesis might be thought of as a building at the high tide mark. Future experiments may wash away the structure, but the foundation will probably remain. Hypotheses requiring the fewest assumptions are the simplest and, therefore, are the best.

A scientific law is a short summary statement, which can often be expressed as a simple algorithm or equation, that states how a group of quantities are related under a set of controlled conditions. One example of a scientific law that can be expressed as an equation is that a given quantity of gas at constant pressure will expand as the temperature increases.

Example 1.3: Which of the following are hypotheses?

a. The number of calories a person consumes will determine his weight.
b. Unicorns are friendly animals.
c. Oranges contain vitamin C.
d. Skiing is fun.

Examples **a** and **c** are testable scientific hypotheses. These statements can be tested. However, remember that we cannot prove a hypothesis, only disprove one. In both of these cases we can design experiments (varying the number of calories, analyzing for vitamin C, etc.) which will test the hypothesis. Other statements may be a matter of opinion, but do not constitute a scientific hypothesis.

Example 1.4. Which of the following are testable hypotheses?

a. Adding polybutylene to our new glue formulation will make the glue stickier.

The characteristic of a hypothesis is that it is testable. This can be tested. Add the polybutylene and measure the result.

b. Somewhere in another universe it is possible to travel faster than the speed of light.
Since we cannot measure travel in another universe, this is not testable.

c. An angel has zero mass.
Due to the shortage of available angels in the lab we cannot test this and therefore it is not a scientific hypothesis.

d. Eating a lot of oat bran will lower blood cholesterol levels.

This can be tested. A group of volunteers could be fed a diet high in oat bran and compared to a control group that was not fed oat bran. We could determine whether or not oat bran had an effect, but we could not prove that oat bran itself lowered cholesterol. If the oat bran group had lower cholesterol, that would be consistent with the hypothesis.

5. Scientific Measurements.

The units of the metric system are given below:

giga	(G)	10^9	one billion
mega	(M)	10^6	one million
*kilo	(k)	10^3	one thousand
deci	(d)	10^{-1}	one tenth
*centi	(c)	10^{-2}	one hundredth
*milli	(m)	10^{-3}	one thousandth
*micro	(μ)	10^{-6}	one millionth
*nano	(n)	10^{-9}	one billionth
pico	(p)	10^{-12}	one trillionth

*These are the most commonly used prefixes.

You should know how to combine the prefixes with the units and be able to write the resulting units. For example, mL is the abbreviation for milliliter and is 1/1000 of a liter or 10^{-3} L. Be careful to remember which prefixes are fractional and which are powers of ten to the positive exponent. For example, a century is a hundred years, but a centigram is 1/100 of a gram.

Example 1.5. Convert each of the following measurements to a unit that replaces the power of ten by a prefix.

a. 7.42×10^{-3} s =

b. 5.41×10^{-6} m =

c. 1.19×10^{-3} g =

d. 5.98×10^{3} m =

a. 7.42 ms (milliseconds); **b.** 5.41 μm (micrometers); **c.** 1.19 mg (milligrams); **d.** 5.98 km (kilometers).

Example 1.6. Express each of the following measurements in terms of an SI base unit.

a. 540 nm

b. 120 kg

c. 5000 km

d. 12 μm

a. First convert nm, 10^{-9} , now add the exponent for converting from 540 to 5.40 (10^{2}) so that the final result is 5.40×10^{-7} m **b.** 1.20×10^{2} kg **c.** 5.000×10^{6} m **d.** 1.2×10^{-5} m.

Example 1.7. Use exponential notation to express each of the following in terms of SI units.

a. 475 nm
b. 225 ns
c. 1415 km
d. 2.26×10^{6} g

First remember the SI units. For length the unit is meter; for time, second; and for mass, kg. **a.** 475 nm = 475×10^{-9} m = 4.75×10^{-7} m ; **b.** 225 ns =225×10^{-9} s = 2.25×10^{-7} s; **c.** 1415 km = 1.415×10^{6} m; **d.** 2.26×10^{6} g = 2.26×10^{3} kg.

6. Definition of temperature and conversion between Celsius, Fahrenheit and Kelvin temperatures

Temperature is difficult to define even though we have an intuitive idea that temperature is a measure of "hotness." Temperature measures the average kinetic energy (energy of motion) of particles in a substance. On the Kelvin scale, absolute zero, or zero Kelvin, is the coldest possible temperature and is the temperature at which molecular motion ceases.

The two temperature scales commonly used by scientists are the Kelvin scale and the Celsius scale. On the Celsius scale water freezes at zero degrees and boils at 100 degrees. The interval between the two temperatures is divided into 100 equal parts, each a Celsius degree. The degree is the same size on the Kelvin scale, but the zero point is different. Therefore, to convert between Kelvin and Celsius degrees, simply use the following equation:

$$t_K = t_C + 273.15$$

The conversion between Celsius and Fahrenheit is a little more complicated because the size of the degree is different on each scale. The freezing point of water on the Fahrenheit scale is 32 °F and the boiling point is 212 °F. The following equation can be used for conversion:

$$t_F = 1.8t_C + 32$$

This equation can be rearranged to convert from Fahrenheit to Celsius. Keep in mind that the Celsius degree is larger than the Fahrenheit degree and therefore t_C must be multiplied by 1.8. The addition of 32 comes from the difference in the zero point for the two scales. The freezing point of water is 32 °F and 0 °C. This should help you keep the conversion in mind. To convert from Fahrenheit to Celsius, rearrange the equation to give:

$t_C = (t_F - 32)/1.8$

Example 1.8. Convert 40 °F to Celsius temperature.

Strategy: First we must rearrange the equation so that the result will yield the temperature in Celsius. Then we can plug in and arrive at the result.
Rearranging the above equation,

$1.8t_C = t_F - 32$

and

$t_C = (t_F - 32)/1.8.$

Now substituting in for t_F

$t_C = (40 - 32)/1.8 = 4.4$ °C.

(For estimation purposes, this is slightly above the freezing point, as is 40 °F.)

Example 1.9. Cesium melts at 28.5 °C. What is the temperature in Kelvins?

The conversion is K = °C + 273.15; therefore,

28.5 °C +273.15 = 301.7 K

Example 1.10. Carry out the following temperature conversions.

a. 100.00 °C to degrees Fahrenheit.

$t_F = 1.8t_C + 32$

$= 1.8(100.00) + 32 = 212.00$ °F

b. 40.0 °C to degrees Fahrenheit

$t_F = 1.8t_C + 32$

$t_F = (1.8 \times 40.0) + 32 = 104.0$ °C

Exercise (text) 1.3.A. Carry out the following temperature conversions.

a. 85.0 °C to degrees Fahrenheit

$t_F = 1.8t_C + 32$

$= 1.8 \times 85.0$ °C + 32

$t_F = 153 + 32 = 185$ °F

b. -12.2 °C to degrees Fahrenheit

$1.8 \times (-12.2) + 32 = 10.0$ °F

c. 355 °F to degrees Celsius

$t_C = (t_F - 32)/1.8$
$= (355 - 32)/1.8 = 179\ °C$

d. -20.8 °F to degrees Celsius
$(-20.8 - 32)/1.8 = -29.3\ °C$

Example 1.11. You are traveling in Canada and begin to feel ill. You measure your body temperature and find it is 38.5 ^{0}C. What is your temperature on the Fahrenheit scale?

$T_F = 1.8(38.5) + 32 = 101.3\ ^0F$ (normal body temperature is approximately 98.6 ^{0}F)

What is normal body temperature in ^{0}C?

$T_C = (98.6 - 32)/ 1.8 = 37.0\ ^0C$

7. Defining the terms precision and accuracy.

Precision of a set of measurements refers to how close individual measurements are to one another. Accuracy refers to how close the measurements are to the "correct" or most probable value.

Example 1.12. A student weighs a sample in lab. In calculating the weight of the sample, he forgets to subtract the weight of the beaker. The student weighs the sample four times and records weights of 10.02 g, 10.01 g, 10.03 g and 10.00 g. Are the measurements of the amount of sample accurate? Are they precise?

The measurements are precise because the weights are within a very narrow range. However, the measurements are not accurate because the weight of the beaker was not subtracted, and the amount of sample is much less than the weight of the beaker and sample combined.

8. Determining the number of significant figures in a numerical calculation

It is helpful to write the following rules in your own words.

1. Significant digits include all nonzero digits.

2. Zeros. There are three classes of zeros:

a. Leading zeros are zeros that precede all nonzero digits. They are not significant because they are merely place holders. For example, 0.0012 has only two significant figures, the zeros are place holders.

b. Zeros between digits. These count as significant figures. For example, 1.0012 has five significant figures.

c. Zeros to the right of the number or "trailing zeros." They are significant only if the number has a decimal point or is an exact number. For example 100 has one significant figure, but 1.00×10^2 has three significant figures.

3. Exact numbers. Numbers determined by counting and not from measurement are exact and can be assumed to have an infinite number of significant figures. For example, 6 eggs. Similarly numbers determined from definitions are also exact. 1 doz = 12 or 2.54 cm = 1 in. These numbers do not limit the significant figures in a calculation.

To express the result of a calculation with the appropriate number of significant figures, apply the following rules: For addition or subtraction, the result is rounded to the same number of decimal places as the term with the fewest number of decimal places. The result of multiplication or division should have as many significant figures as the factor with the fewest number of significant figures has.

Example 1.13. Determine the number of significant figures in the following calculation: 1.02 + 2.30 + 0.1 = ?

The limiting term here is 0.1 because it has the fewest decimal places. The final answer will have only one place to the right of the decimal point. The answer will be 3.4.

Example 1.14. Determine the number of significant figures in the calculation: (2.560 x 8.4)/(1.24) = ?

Carrying out the calculation will give 17.341935. However, 8.4 has only two significant figures, so the number of significant figures in the answer is two and the correct answer is 17. With multiplication and division, the number of significant figures is what is important, not the location with respect to the decimal point. Notice that rounding for significant figures is done **after** all the calculations. In many cases a different (and incorrect answer) will be obtained if rounding is done for intermediate terms.

Example 1.15. Perform the indicated operations and give answers with the proper number of significant figures.

a. 73 m x 1.340 m x 0.41 m

The answer from the calculator is 40.1062. The correct number of significant figures is two because that is the number of significant figures in the factor with the least significant figures; in this case 73 m and 0.41 m both have two significant figures. The answer is 40. m^3 or 4.0 x 10^1 m^3.

b. 0.137 cm x 1.43 cm = 0.196 cm^2, three significant figures. Remember that for multiplication and division we count the number of significant figures in the factor with the least significant figures.

c. 3.132 cm x 5.4 cm x 5.4 cm

= 91 cm^3, two significant figures

d. 51.79 m/4.6 s

= 11 m/s

e. 456.1 mi/7.13 h

= 64.0 mi/h

f. 305.5 mi/14.7 gal
= 20.8 mi/gal

9. Use the Unit-Conversion method for solving problems that require numerical calculations.

You use the unit conversion method in solving many everyday problems, but you are probably not familiar with exactly how you do this and have not written the appropriate equations. Take the following example: a recipe for sugar cookies requires 3 eggs and makes one dozen cookies. You wish to make 4 dozen cookies. How many eggs do you need to buy? The first part of this can be written algebraically as:

1 recipe = 1 doz cookies, 4 doz x 1 recipe/doz = 4 x recipe.

Since each recipe requires 3 eggs, 4 recipe x 3 eggs/recipe = 12 eggs and 12 eggs x 1 doz/12 eggs = 1 doz eggs. You probably solved this problem in your head without thinking about the steps involved, which are just a series of unit conversions.

Know the common units of the English system for volume, length, and mass, and the relationships between them.

Volume

Gallon (gal)	= 4 quarts (qt)	
Quart (qt)	= 2 pints (pt)	
Pint (pt)	= 2 cups	= 16 fluid ounces (fl oz)

Length

Mile (mi)	= 1760 yards (yd)	= 5280 feet (ft)
Yard (yd)	= 3 feet (ft)	= 36 inches (in)

Mass

Ton (t)	= 2000 pounds (lb or #)
Pound (lb)	= 16 ounces (oz)

If you are familiar with the English and metric units given above, conversion is a simple process. You need remember only one conversion for length, one for volume, and one for mass. Conversion then becomes an algebraic exercise, and if you pay attention to units, it should pose no great difficulty. Given below are several conversion factors, but remember you need only learn one from each column.

Mass	Length	Volume
28.35 g = 1 oz	2.54 cm = 1 in	29.6 mL = fl oz
453.6 g = 1 lb	39.37 in = 1 m	0.9464 L = qt
kg = 2.205 lb	30.48 cm = ft	

Example 1.16. A baseball pitcher's fast ball is clocked by radar at 98 mi/h. What is the speed in meters per second?

Strategy: This is a unit conversion problem. How we go about solving it depends to some degree on which unit conversions we are most confident of remembering. Since we are very familiar with the English system, and the easiest metric conversion for us to remember is 2.54 cm = 1 in, we will first convert from mi to in and then to the metric system. If you have other conversions, such as mi to km, you can shorten the process, but it is not necessary to learn all of these and we will get to the same answer. Therefore, the first conversion is to in, in to cm, and then time is converted from hr to min to sec.

$$\text{speed} = 98\frac{\text{mi}}{\text{h}} \text{ x } \frac{5280\text{ ft}}{\text{mi}} \text{ x } \frac{12\text{ in}}{\text{ft}} \text{ x } \frac{2.54\text{cm}}{\text{in}} \text{ x } \frac{1\text{ m}}{100\text{cm}} \text{ x } \frac{1\text{ h}}{60\text{ min}} \text{ x } \frac{1\text{ min}}{60\text{sec}} = 44\text{ m/s}$$

Example 1.17. Carry out the following conversions.

a. 76.3 mm to m

$$76.3\text{ mm x } \frac{1\text{ m}}{10^3\text{mm}} = 0.0763\text{ m}$$

b. 0.0856 kg to g

$$0.0856\text{ kg x } \frac{10^3\text{g}}{\text{kg}} = 85.6\text{ g}$$

c. 0.927 lb to oz

$$0.927\text{ lb x } \frac{16\text{oz}}{\text{lb}} = 14.8\text{ oz}$$

d. 415 in. to yd

$$415\text{ in. x } \frac{1\text{ft}}{12\text{ in}} \text{ x } \frac{\text{yd}}{3\text{ ft}} = 11.5\text{ yd}$$

e. 3.00 L to fl oz

$$3.00\text{ L x } \frac{10^3\text{mL}}{\text{L}} \text{ x } \frac{1\text{ fl oz}}{29.57\text{mL}} = 101\text{ fl oz}$$

Example 1.18. Carry out the following conversions.

a. 90.0 km/h to m/s

$$\frac{90.0\text{ km}}{\text{h}} \text{ x } \frac{10^3\text{ m}}{\text{km}} \text{ x } \frac{\text{h}}{60\text{min}} \text{ x } \frac{\text{min}}{60\text{s}} = 25.0\text{ m/s}$$

b. 1.39 ft/s to km/h

$$\frac{1.39\text{ft}}{\text{s}} \text{ x } \frac{12\text{in}}{\text{ft}} \text{ x } \frac{2.54\text{cm}}{\text{in}} \text{ x } \frac{\text{m}}{100\text{ cm}} \text{ x } \frac{\text{km}}{1000\text{ m}} \text{ x } \frac{60\text{s}}{\text{min}} \text{ x } \frac{60\text{ min}}{\text{h}} = 1.53\text{ km/h}$$

c. 4.17 g/s to kg/h

$$\frac{4.17\text{g}}{\text{s}} \text{ x } \frac{\text{kg}}{1000\text{g}} \text{ x } \frac{60\text{s}}{\text{min}} \text{ x } \frac{60\text{ min}}{\text{h}} = 15.0\text{ kg/h}$$

Example 1.19. Carry out the following conversions.

a. 476 cm^2 to in.^2

$$476\ \text{cm}^2 \times \frac{\text{in}}{2.54\text{cm}} \times \frac{\text{in}}{2.54\text{cm}} = 73.8\ \text{in.}^2$$

Note that we use the conversion factor twice to cancel the cm^2 term. This is an important aspect of dimensional analysis involving known unit factors.

b. 124 ft^3 to m^3

$$124\ \text{ft}^3 \times \left(\frac{12\text{in}}{\text{ft}}\right)^3 \times \left(\frac{2.54\text{cm}}{\text{in}}\right)^3 \times \left(\frac{\text{m}}{1000\text{cm}}\right)^3 = 3.51\ \text{m}^3$$

c. 15.8 lb/in.^2 to kg/m^2

$$\frac{15.8\ \text{lb}}{\text{in}^2} \times \frac{\text{kg}}{2.205\text{lb}} \times \left(\frac{\text{in}}{2.54\text{cm}}\right)^2 \times \left(\frac{100\text{cm}}{\text{m}}\right)^2 = 1.11 \times 10^4\ \text{kg/m}^2$$

Example 1.20. Estimate the following quantities. Do not do a detailed calculation, but learn to "guess" the amounts. This will give you an idea of some of the metric units you may not be familiar with.

a. your weight in kg
b. your height in meters
c. the volume of a glass of milk in mL
d. 60 mi/hr in km/hr
e. the distance in mi of a 5 km race.

a. A kilogram is a little over 2 lbs. Therefore if you weigh 150 lbs, you weigh approximately 75 kg. (An actual calculation will show that you weigh 68 kg. The estimate is fairly close and very useful in determining whether your answers are reasonable when solving problems.)
b. A meter is about 3 ft. If you are 6 ft tall, you are approximately 2 m tall.
c. A quart of milk is about a liter. If your glass contains a pint, then it contains one-half a quart or approximately 500 mL, one-half of a liter.
d. 1 mi is approximately 1.5 km. 60 mi/hr is approximately 90 km/hr. These estimates will also help you so that you do not confuse the conversion factors. A common mistake is to invert the conversion factor. If you do these simple exercises, it will help you to remember that 1 km = 0.6214 mi and not the reverse!
e. 1 mi is about 1.5 km, so a 5-km race is about 3.2 mi.

Density can be used as a conversion factor to convert between the mass and volume of a substance. A common sense approach says that a dense object is very heavy for its size. A bag of lead balls is heavier than a similar size bag of feathers. In more precise terms we are saying that the mass/volume ratio (density) is greater for lead than for feathers. Density is a physical property and is defined as the mass/volume. Units of density will be mass/volume (g/mL).

Example 1.21. A 1.00 L flask of liquid has a mass of 2.00 lb. What is the density of the liquid in g/mL?

Strategy: This is a two part problem since we must first calculate density and then convert to the appropriate units.

Calculation: Density is mass/volume so the density is 2.00 lb/1.00 L = 2.00 lb/L. To convert to g/mL we use the above conversion factors:

$$\frac{2.00\text{ lb}}{\text{L}} \times \frac{453.6\text{g}}{\text{lb}} \times \frac{\text{L}}{1000\text{ mL}} = 0.907\text{ g/mL}$$

Notice that 1 lb is about 500 g and therefore our answer should come out to be about 1 g/mL. Estimating answers will prevent you from making many errors in conversions. Think about the relative size of units and whether or not the answer makes sense. A common density for many aqueous and liquid solutions is about 1 g/mL.

Example 1.22. Calculate the density of a solution that has a mass of 40.7 g and occupies a volume of 38.9 mL.

Density is mass/unit volume. Therefore we divide 40.7 g by 38.9 mL. 40.7 g/38.9 mL= 1.0462725. The correct number of significant figures is three. So the answer is rounded to 1.05 g/mL.

Example 1.23. Calculate the mass of 40.0 mL of ethanol. The density of ethanol is 0.789 g/mL.

Here we want to know the mass of a given volume. Density tells us the mass/unit volume, so this is a simple multiplication. 40.0 mL x 0.789 g/mL = 31.56 g and rounding to the correct number of significant figures (three) gives 31.6 g. Since 1 mL has a mass of less than 1 g, 40 mL should be less than 40 g and this answer seems reasonable.

Example 1.24. An experiment calls for 8.65 g of carbon tetrachloride, the density of which is 1.60 g/mL. What volume would you take?

Here we know the mass and need to find the volume. Density gives us mass/volume. If we divide the mass by the density, we will have the volume. A common example of this is if you want 8 oz of chocolate to bake a cake and each package contains 2 oz. You divide the amount you want (8 oz) by the amount/package (2 oz). You will need four packages. The same applies in dealing with g/mL. We need 8.65 g and each mL contains 1.60 g.

Therefore, we divide:

$$8.65\text{ g} \times \frac{\text{mL}}{1.60\text{ g}} = 5.41\text{ mL}$$

Density can be thought of as another pair of unit conversion factors for converting volume to grams or grams to mL.

Additional Note: All conversion factors are simply using algebra. If the density is 1.60 g/mL, then 1.60 g = 1 mL using simple algebra, the conversion factor of 1.60 g/1 mL can be multiplied or divided to give the appropriate units.

10. Practice estimation exercises to determine if answers are reasonable.

For conversion exercises it may be useful to think of common objects and their approximate metric length, mass or volume (See Example 1.15). Several examples of estimation are given in the above problems.

Example 1.25. Estimate the answers to the following problems:

a. 1.02 x 0.096 x 31 =
b. 4.112 – 1.008 =
c. (2.312 x 1.899) – (4.211) =
d. (9.976 x 4.87)/(1.22 + 5.34) =

a. This can be estimated as 1 x 0.1 x 30 = 3
b. 4 – 1 = 3
c. (2 x 2) – 4 = 0
d. (10 x 5) / 6 = 8

All of these answers are estimates, but if you are doing a complicated calculation, you will be able to judge whether your answer is reasonable. It works very well for exams and in the laboratory to determine whether you have entered the numbers correctly on your calculator. The earlier example of estimating common quantities will help you even more to determine if your answers are reasonable.

Skills Test Problem

The concentration of lead (Pb) in highly polluted air was determined to be 5.0×10^{-6} g Pb/m^3 air. An average male breathes about 8.20×10^3 L of air per day. If 30% of the Pb inhaled is retained in the lungs, what is the mass of lead absorbed from polluted air in one year?

Solution

Examining the information given, we must first convert the concentration units (g/m^3) to match the volume (L). It doesn't really matter if m^3 is converted to L or vice versa. For our example, we will convert to L.

$$\frac{5.0\times10^{-6}\ \text{g}}{\text{m}^3}\times\frac{1\ \text{m}^3}{(100\ \text{cm})^3}\times\frac{1\ \text{cm}^3}{\text{mL}}\times\frac{1000\ \text{mL}}{\text{L}} = \frac{5.0\times10^{-6}\ \text{g Pb}}{1.0\times10^3\ \text{L}} = 5.0\times10^{-9}\ \text{g Pb / L}$$

Next we calculate how much Pb is inhaled in one day and in one year.

5.0×10^{-9} g Pb/L x 8.20×10^3 L/day x 365 days/ 1 yr = 1.5×10^{-2} g/yr

Assuming 30 % of the lead is retained, then the mass is $0.30 \times 1.5 \times 10^{-2}$ g/yr = 4.5×10^{-3} g/yr or 4.5 mg/yr.

Quizzes

The following short quizzes are designed to help you test your mastery of each subject. Limit yourself to about 15 min for each quiz. Answers will be found at the end of this manual.

Quiz A

1. How many micrograms are there in a kilogram?
2. How many significant figures are there in the result of the following calculation?
   $(7.13 \times 10^{-5}) \times (1.20 \times 10^{2}) + 0.00345 = ?$
3. A picture frame measures 4.0 x 6.0 inches. What is the area of the picture frame in square centimeters?
4. Matter which cannot be further broken down by chemical means is called a(n)

   a. element

   b. compound

   c. mixture

   d. heterogeneous mixture

   e. none of these.

5. The density of a solid was found to be 1.354 g/cm^3. What volume does 2.54 g of the solid occupy?
6. The temperature on a cold winter day is about

   a. 3 °C b. 40 °C c. 60 °C d. 400 °C e. 10 K.

7. A container of Italian salad dressing is an example of

   a. an element

   b. a homogeneous mixture

   c. a compound

   d. a heterogeneous mixture

   e. none of the above.

8. A scientific hypothesis

   a. is always right

   b. can be proven

   c. can be tested

   d. results from many experiments

   e. all of the above.

9. Convert 72 °F to Celsius and Kelvin temperatures.

Quiz B

1. How many significant figures are in each of the following numbers?
   a. 1.00456
   b. 45,000
   c. 0.00230
2. Which of the following is a physical change?
   a. an animal produces fat from eating
   b. carbon dioxide freezes to form dry ice, solid carbon dioxide
   c. water is decomposed to form oxygen and hydrogen gas
   d. none of these.
3. An explanation of observations which has been proven true by many tests is called
   a. a hypothesis
   b. a homogeneous mixture
   c. a law
   d. an experiment
   e. none of these.
4. Carry out the following conversions:
   a. 24.9 °C to K
   b. 30 °C to °F
5. Change the following units to ones with appropriate metric prefixes:
   a. 1.23×10^{-6} g
   b. 1230 mL
   c. 4.78×10^{6} m
6. Convert the following numbers to the units indicated:
   a. 42.0 in to m
   b. 12.2 mL to oz
   c. 7.00 g to lb

Quiz C

True or False

1. The more times a measurement is made with an appropriately standardized instrument, the more accurate the measurement will be.
2. A warm day in Atlanta, GA has a temperature of 15 °C.
3. When ice melts it is an example of a physical change.
4. Orange juice is an example of a heterogeneous mixture.

Give answers to the following questions using the correct number of significant figures where appropriate:

5. 107.334 + 1.23 +379.6 =
6. The density of plutonium (Pu) is 19.8 g/cm^3. What is the volume of 10 lbs of Pu?
7. Convert 10 mL to liters.
8. A runner can run a mile in 4.0 min. Assuming he can consistently run at the same speed, how long will it take him to run 10,000 meters?
9. Convert 100 ng to mg.
10. Convert 90 °C to degrees Fahrenheit and degrees Kelvin.

Self Test

1. Which of the following statements describe physical changes and which describe chemical changes?
   a. milk going sour
   b. slicing cheese
   c. drying clothes
   d. burning of wood
   e. forming plastic into a mold
2. Classify each of the following as a pure substance, heterogeneous mixture, or homogeneous mixture.
   a. raisin bread
   b. tap water
   c. lemonade
   d. dry ice
   e. sulfur
   f. sugar
3. Convert
   a. 102 cm to mm
   b. 450 L into mL
   c. 3 ft into cm
   d. 15 joules into kilojoules
   e. 15 m/s into mi/hr

Multiple Choice

4. If 3.09321 were divided by 0.0142, the answer would contain
   a. six significant figures
   b. five significant figures
   c. four significant figures
   d. three significant figures

e. two significant figures

5. A physical property is one that
   a. changes the composition of the material
   b. does not depend on the size of the sample
   c. can be determined without changing the composition of the material
   d. cannot be described
   e. must be measured many times

6. The following data were collected by a student in chemistry laboratory:
   15.78 mL, 15.65 mL, 15.90 mL

   The true volume for the experiment is 21.0 mL. The measured data is:
   a. very accurate but not precise
   b. very precise but not accurate
   c. very accurate and very precise
   d. neither accurate nor precise

7. Density can be used as a conversion factor to relate
   a. grams to pounds
   b. mass to volume
   c. mass to hardness
   d. volume to fluidity
   e. mass to composition

8. **Matching**

Hypothesis	a. the amount of space occupied by an object
Accuracy	b. a fundamental substance that can't be chemically changed or broken down
Density	c. an interpretation that explains the results of many experiments
Element	d. a consistent explanation of known experiments
Precision	e. the mass of a given volume of sample
Volume	f. how close a measurement is to the true value
Theory	g. how close a set of measurements are to each other

Problems

9. A basketball player is 7 ft tall. What is his height in m?
10. A burglar is about to steal a gold statue from a museum. The statue is on a weight sensitive platform that will sound an alarm if the weight is removed. The thief wishes to replace the statue with a bottle of water. The volume of the statue is 500 mL. What volume of water should he use? (The density of gold is 19.3 g/cm^3 and the density of water is 1.0 g/cm^3)

11. A student exercises on a treadmill and burns 530 cal/hr. The student exercises for 25 min. How many hot dogs, which contain 160 calories each, must the student eat to provide the energy burned on the treadmill?
12. The boiling point of ethanol is 78.5 °C. Convert this temperature to °F.
13. At what temperature will the reading be the same on a Celsius and Fahrenheit thermometer?
14. What is the weight of 1.000 gallon of water? Would you expect the weight of a gallon of ice to be more or less?
15. A car gets 35 mi/gal of gasoline. Gas costs $2.00/gal. How much gas is required to drive 500 mi and how much does it cost?
16. Which is a bigger change in temperature, 10 °C or 10 °F?
17. A typical blood cholesterol reading is 180 mg/dL. What is the reading in g/L?

CHAPTER 2

Atoms, Molecules and Ions

This chapter is devoted to helping you understand the symbols and relationships of chemical reactions. Much of this chapter deals with the language of chemistry: how chemists express reactions and compounds in ways that are understood by all chemists. We will also explore some of the underlying natural laws and principles that underlie our understanding of the atoms, molecules and ions that make up matter.

Chapter Objectives:

You should be able to :

1. State and understand the laws of conservation of mass and definite proportions.
2. State the basic assumptions of Dalton's atomic theory.
3. Know the subatomic particles: neutrons, protons and electrons and their important properties.
4. Define isotopes and know the number of neutrons, protons and electrons from the mass numbers.
5. Define the terms atomic mass unit and atomic weight.
6. Calculate the atomic mass of an element from the isotopic abundance and calculate the isotopic abundance from the atomic mass.
7. Distinguish metals and nonmetals from the periodic table.
8. Write empirical and molecular formulas.
9. Know the names and how to write the formulas for binary compounds.
10. Define ions and ionic substances and be able to write the formulas for ions.
11. Know the formulas of polyatomic ions and how to name compounds containing polyatomic ions.
12. Know the characteristics of acids, bases and salts, according to the Arrhenius concept
13. Name simple acids and bases.
14. Name simple organic compounds and know organic functional groups.

Chapter Summary

1. Laws of conservation of mass and definite proportions.

Explain in your own words the law of conservation of mass: The total mass of the products of a reaction is always equal to the total mass of the reactants (starting materials) consumed in the reaction.

Example 2.1. A sample containing 5.00 g of methane was burned, using 20.0 g of oxygen and producing 11.25 g of water. How many grams of carbon dioxide were produced?

Solution: The sum of the masses of the reactants (methane and oxygen) must equal the sum of the masses of the products (water and carbon dioxide). Therefore,

mass of methane + mass of oxygen = mass of water + mass of carbon dioxide
5.00 g methane + 20.0 g oxygen = 11.25 g water + mass carbon dioxide
25.00 g = 11.25 g + mass carbon dioxide
The mass of carbon dioxide is 13.75 g.

The law of definite proportions states that a particular compound always contains its constituent elements in certain fixed proportions and in no other combinations. An element will of course form more than one compound, but for a given compound, the elements will always have the same proportions. The law as stated talks about proportions by mass. The concept of atomic ratios results from atomic theory, which is derived from these empirical laws. For example, water always contains two hydrogen atoms to one oxygen atom.

Example 2.2. Iron (II) sulfide contains 36.47% sulfur by mass. If 87.91 g of iron sulfide is made by reacting iron with sulfide, what mass of iron was reacted?

Solution: By the law of definite proportions, the iron (II) sulfide produced contains 36.47% by mass sulfur and 63.53 % by mass iron (100% - % of sulfur). The mass of iron in 87.91 g iron sulfide is (63.53 % x 87.91 g)/100 % = 55.85 g. The fraction of iron (II) in iron (II) sulfide is more than half, so by estimation, 55.84 g is a reasonable answer. If you forgot to divide by 100% you will notice you have more iron than total sample!

2. Basic assumptions of Dalton's atomic theory.

You should restate these in your own words. Brief versions of Dalton's assumptions are: (a) Elements are composed of indestructible atoms. (b) Different elements have different atoms. (c) Compounds are formed when atoms combine in small whole number ratios. Dalton also postulated that all atoms of a given element are identical. This was incorrect, he did not picture isotopes.

Example 2.3. A nitrogen-oxygen compound is found to have 1.142 g oxygen per gram of nitrogen. Which of the following oxygen-to-nitrogen mass ratio(s) is(are) also possible for a nitrogen-oxygen compound? (a) 0.612 (b) 1.250 (c) 1.713 (d) 2.856.

Solution: To calculate the possible ratios, we must determine which have small whole number multiples of the given mass ratio. We know that 1.142 is an acceptable ratio from the information given. We want to determine whether any of the ratios are simple multiples of this ratio. We could write this as (1.142) x (X) = new ratio, then divide the new ratio by 1.142 and determine whether X or a simple multiple of X is a small whole number. For ratio (a) 0.612:1.142 = 0.536. In terms of ratio this is 0.536:1 and to get whole numbers would be 536:1000, not a small whole number. Ratio (b) 1.250:1.142 = 1.095. Again this will not give a small whole number ratio. For (c) 1.713:1.142 = 1.500. This gives 1.5:1 or 3:2 ratio and this is a ratio of small whole numbers. (d) 2.856:1.142 = 2.501. This ratio is 5.002:2. We accept this as 5:2 and this is an acceptable ratio of small whole numbers.

3. Subatomic particles: protons, neutrons, and electrons and their important properties.

Protons are subatomic particles with an approximate fundamental mass of one and a charge of +1. They are part of the nucleus. Neutrons also have a fundamental mass of one, but have no charge. The electron has a much smaller mass (1/1836 of the proton), a charge of -1 and is found outside the nucleus. Protons and neutrons are found in the nucleus, while electrons are outside the nucleus. The number of protons is the same for isotopes of a given element.

4. Isotopes and mass number.

The integer mass number of an atom of a certain isotope is the sum of the number of neutrons and protons in the atom nucleus. Isotopes are atoms that have the same number of protons, but different numbers of neutrons. Examples are hydrogen (protium), deuterium, and tritium. Many other examples occur in nature.

Notice that the number of protons does not change for isotopes of the same element. We will see examples of different numbers of neutrons (isotopes) and electrons (ions), but protons are the same for a given element.

Example 2.4. Consider the following table:

Neutrons	Protons	Electrons
8	6	6
7	6	6
7	7	7
7	7	8
6	6	6

Which of these are isotopes of carbon? Which of these species are ions?

Solution: From the above statement, isotopes of carbon will all have six protons. Therefore, the first two lines and last line of the table will represent isotopes of carbon. The first element will be carbon-14 and will not have a net charge. Similarly the second element is carbon-13, again neutral charge. The last element is carbon 12 with neutral charge. Lines 3 and 4 represent nitrogen-14. The third line represents nitrogen with a neutral charge and the fourth line represents nitrogen with a -1 charge.

The number of protons, neutrons and electrons present in atoms and ions are designated using the symbolism ${}^{A}_{Z}X$. The number in the upper left, A, is the mass number. This is the sum of the number of neutrons and protons in the nucleus. The electrons, because of their much smaller mass, are not counted. (They are also not in the nucleus.) Isotopes are often written ${}^{32}P$ or sometimes P-32.

The subscript in the lower left, Z, is the atomic number and represents the number of protons in the nucleus. The atomic number is actually unnecessary since the elemental symbol determines the atomic number. This number is often included for emphasis and in writing radioactive decay equations.

If a superscript appears in the upper right it represents the charge of the atom. The charge is the number of protons minus the number of electrons. The sign of the upper right superscript is always included and can be either positive or negative. The charge of a neutral atom is always zero and this superscript is assumed unless otherwise given.

Example 2.5. What are the number of protons, neutrons and electrons in $^{35}Cl^{-1}$?

Solution: The mass number is 35, the number of protons + neutrons. The atomic number of Cl is 17 (this is given on the periodic table). The number of protons is therefore 17 and the number of neutrons is 35 - 17 = 18. The charge is -1, the number of protons - electrons. The number of electrons is given by 17(number of protons) - electrons = -1 and the number of electrons is 18.

Example 2.6. What is the mass number of an isotope of cobalt that has 33 neutrons?

Solution: The atomic number of cobalt is 27. The mass number is 27 (the number of protons) + 33 = 60. This is often referred to as cobalt-60.

5. Atomic mass unit and atomic mass.

The atomic mass unit is 1/12 the mass of a carbon-12 atom. The atomic mass listed for an element represents a number-average of the actual isotopic masses of a suitable sample of that element. This is a decimal number given in the periodic table and represents all the isotopes of the element.

Looking ahead to Section 6.5, we can calculate the atomic masses. Atomic masses are calculated by using the abundance of each isotope. Let's consider an example to make this type of calculation clear. Suppose you are buying CDs at \$11.00/disc and some tapes at \$5.00/tape. You are buying four discs and three tapes. The total cost will be

\$11.00 x 4 = \$44.00
\$ 5.00 x 3 = \$15.00
Total= \$59.00

The average cost of an item will be \$59.00/7 = \$8.43. Another way to calculate this is

$$\left(\frac{4 \text{ discs}}{7 \text{ items}} \times \frac{\$11.00}{\text{disc}}\right) + \left(\frac{3 \text{ tapes}}{7 \text{ items}} \times \frac{\$5.00}{\text{tape}}\right) = \$8.43/\text{item}$$

The fraction of each item of a given price is multiplied by the price and the sum taken. This is the same type of calculation used to calculate atomic masses. For elements, the fractional abundance of each isotope is multiplied by the atomic mass of the isotope and the masses are then summed.

Example 2.7. Bromine consists of two isotopes with masses of 78.92 u and 80.92 u . The average atomic mass of Br is 79.904. What is the relative abundance of each isotope?

Solution: The atomic mass of Br is determined by

Atomic mass = ((% abundance isotope 1) x (mass isotope 1) + (% abundance isotope 2) x (mass isotope 2))/100 or if we use fractional abundance, we can eliminate the division by 100.

Since there are only two isotopes, the sum of the fractional abundance must be 1.0. Therefore, we can rewrite the equation:

Atomic mass = (1- abundance of isotope 2) x (mass isotope 1) + (abundance of isotope 2) x (mass isotope 2)

Remember that these are fractional abundance (%abundance/100).
If x = fractional abundance of isotope 2, and substituting into the equation:

$$79.904 = (1-x)(78.92) + (x)(80.92)$$
$$79.904 = 78.92 - 78.92x + 80.92x$$
$$79.904 = 78.92 + 2.00x$$
$$0.984 = 2.00x$$
$$x = 0.4920$$

or 49.20% of the Br is ^{81}Br (atomic mass 80.92). If we look at the masses of the two isotopes and the atomic mass, this is plausible since the atomic mass of Br is approximately the average of the mass of the two isotopes and one would expect approximately equal abundancies.

7. Metals and nonmetals using the periodic table.

The periodic table lists the known elements arranged in order of increasing atomic number. Groups are vertical columns of elements having similar properties. Periods are the horizontal rows of elements.

Elements to the left of the heavy black line (except hydrogen) are metals. Elements to the right of the stepped heavy line are nonmetals.

Example 2.8. Classify the following compounds as metals or nonmetals.

a. Na **b.** P **c.** Ba **d.** Ar

Solution. By looking at the periodic table, we can determine whether these are metals or nonmetals. Elements to the left are metals and those to the right are nonmetals.

a. metal
b. nonmetal
c. metal
d. nonmetal

8. Writing chemical formulas.

A chemical formula represents the composition of a compound. The formula will indicate both the elements present and the relative numbers of each element. Common table salt is NaCl. This indicates one atom of sodium is present for each atom of Cl. For the more complex formula NaH_2PO_4, this shows that sodium, Na, hydrogen, H, phosphorus, P and oxygen, O are present in the ratio 1:2:1:4, respectively. That is, one atom of sodium, two atoms of hydrogen, one atom of phosphorus and four atoms of oxygen. Note that the chemical symbols are used for the elements, and the subscripts denote the relative number of atoms present.

A **structural formula** shows how the atoms are connected to one another.

An **empirical formula** is the simplest formula that represents the correct ratio of elements. The **molecular formula** represents the exact number of atoms in the compound and must be a multiple of the empirical formula. In many cases, the empirical and molecular formulas will be the same. For example, methane, CH_4, has the same empirical and molecular formulas. A difference between empirical and molecular formulas is most commonly encountered in organic compounds containing carbon. For example, CH_2O is the empirical formula for glucose ($C_6H_{12}O_6$), a type of sugar, and acetic acid ($C_2H_4O_2$), found in vinegar.

Example 2.9. The empirical formula for a hydrocarbon is CH_2. Its atomic mass is 84 u. What is the formula for the compound?

Solution: We know that the molar mass must be a whole number multiple of the empirical mass since the formula of the compound is a multiple of the empirical formula.

The formula mass of CH_2 is 14.0 u. Dividing 84/14 = 6. The molecular formula is 6 x CH_2 or C_6H_{12}, the formula for cyclohexane and a number of other isomers.

9. Names and formulas for binary molecular compounds.

Binary compounds are composed of two elements. Prefixes mono-, di-, tri-, tetra- penta-, etc. are used to indicate the number of atoms of an element in the formula. The first element name is left unchanged and the second element name ending is changed to -ide. Sometimes these designations are understood, as we will see later with ionic compounds.

Example 2.10. Name the following compounds:

PCl_5, PCl_3, CO, CO_2, N_2O

Solution: The compound PCl_5 contains one phosphorus and five chlorine atoms. The name is phosphorus (leave the name of the first element unchanged) pentachloride (penta indicates five and the element chlorine is changed to an -ide ending).

Phosphorus trichloride is the name for PCl_3.

CO is carbon monoxide. Notice that we drop the second –o- in mono so that the compound is carbon monoxide, not carbon monooxide. This is simply to make pronunciation easier.

CO_2 is carbon dioxide. Again, the prefix di- indicates two and the ending of oxygen is changed to -ide.

N_2O is dinitrogen monoxide. We need to indicate that two nitrogen atoms and one oxygen atom are present.

Example 2.11. Write the formulas for the following compounds:

nitrogen dioxide
dinitrogen tetroxide
sulfur dichloride

Solution: Nitrogen dioxide indicates that the first element is nitrogen, N. The second element is oxygen and the prefix di indicates that two oxygen for each nitrogen. The formula is therefore NO_2.

Dinitrogen tetroxide indicates again that the first element is nitrogen, but di- indicates two. The tetroxide indicates four oxygen. The formula is therefore N_2O_4. (What would be the empirical formula for this compound?)

Sulfur dichloride has sulfur as the first element and chlorine as the second. Di- indicates two chlorine and the formula is SCl_2.

Example 2.12. What is the name for S_4N_2?

Solution: The first part by now is easy: tetrasulfur. The second part requires changing the nitrogen to an- ide ending: nitride. The compound is tetrasulfur dinitride (four sulfur, two nitrogen).

10. Ions and ionic compounds.

An ion is a charged particle composed of one or more atoms. For a single atom, this means that the number of electrons is not equal to the number of protons. An ion can have either a positive or a negative charge. Ions of simple atoms are formed when an atom gives up or gains one or more electrons. The periodic table will help us predict simple ions. Group 1A atoms give up one electron to form ions with a +1 charge. Group 2A give up two electrons to form +2 ions. For metal atoms in the A Group, the number of electrons given up is often equal to the group number. The nonmetals of Group 7A take up one electron to form ions with a -1 charge and those of Group 6A take up two electrons to form ions with a -2 charge.

Binary (two atoms) simple ionic compounds are made up of anions (negatively charged) and cations (positively charged) to form a neutral compound. These are usually composed of a metal (cation) and a nonmetal (anion).

Example 2.13. Determine the formula unit for compounds formed from the following pairs of elements.

a. sodium and chlorine

b. calcium and fluorine

c. magnesium and bromine

d. aluminum and sulfur

Solution: The first step is to write the ions involved in forming these compounds. For example **a,** sodium is a Group 1A element and, therefore, its cation has a +1 charge. Chlorine belongs to group 7A and its anion has a -1 charge.

The second step is to combine these ions to form a neutral compound. In this case one sodium ion can combine with one chloride ion to form a neutral compound NaCl. Ionic compounds are named by using the name of the first element (the positive ion) and changing the ending of the second element to ide. NaCl is sodium chloride.

Example **b.** Calcium is a Group 2A element and will have a +2 charged ion. Fluorine belongs to Group 7A and its ion has a -1 charge. Combining one ion of calcium with one of fluoride will give a net charge of +1, which will not work since compounds must be neutral. We need to use two fluoride ions to each calcium ion, and the formula will be CaF_2. We have made sure that the total charge is zero. For more complicated examples the solution may not be so obvious at first; however, if we multiply the charge of each ion by the charge of the opposite ion we will have the number of ions necessary for a neutral atom. For example, suppose the positive ion has a charge of +2 and the negative ion has a charge of -3. Multiplying the positive ion by 3 gives a total positive charge of +6 and multiplying the negative ion by 2 gives a total negative charge of -6. The compound will be neutral and written as X_3Y_2. We would then look to see if the multiples of the ions are the simplest possible. They must be whole numbers. For example, if the cation, X^{4+}, has a charge of +4 and the anion, Y^{2-}, has a charge of –2, we would do the following: using our multiplication technique we would multiply X by 2 for a positive charge of +8 and Y by 4 for a negative charge of -8. The charges balance and we would have a compound of X_2Y_4. However, 2 and 4 can each be divided by 2 to give the formula XY_2. This is still neutral with a charge from the positive ion of +4 and the negative ion of -2 x 2 = -4.

Applying these techniques, to examples **c** and **d** we have $MgBr_2$, magnesium bromide, and for example **d**, we have Al^{3+} and S^{2-}, which is written as Al_2S_3, aluminum sulfide. Since there is no simple whole number to divide 2 and 3, this is the correct formula. Notice that two important things are involved: You must know the charge of the ions (this can usually be determined by the periodic table for most ionic compounds) and the compound must be neutral.

11. Polyatomic ions and naming compounds containing polyatomic ions.

The same principles described above apply to writing formulas for polyatomic ions. The polyatomic ions are considered the same way the simple ions are and the charge is associated with the group of atoms that make up the polyion. A number of these are given in Table 2.4 of the text. It is important to memorize the most common polyatomic ions.

Example 2.14. Give names or formulas of the following compounds:

a. ammonium nitrate b. $FePO_4$

Solution: The ammonium ion is NH_4^+ and the nitrate ion is NO_3^-. Since we have one positive and one negative charge the compound is written NH_4NO_3.

b. $FePO_4$ iron (III) phosphate. Use the name of the positive ion or element first and then the negative ion, phosphate, PO_4^{-3}. The Roman numeral III indicates Fe has a +3 charge. This must be the case to give a neutral compound.

Using the prefixes hypo- and per- and the suffixes -ite and -ate.
These prefixes and suffixes refer to ions with increasing number of oxygen atoms:
hypo____ite < ______ite < ______ate < per______ate

For example, ClO^-, hypochlorite ion < ClO_2^-, chlorite ion < ClO_3^-, chlorate ion < ClO_4^-, perchlorate ion.

12. Characteristics of acids, bases and salts, according to the Arrhenius concept.

Simple acids are those that produce H^+, hydrogen ions, when dissolved in water. Acids have a sour taste, turn litmus paper from blue to red and react with active metals to dissolve the metals and produce hydrogen gas. Acids neutralize bases, and usually dissolve metal oxides forming water and salts containing metal cations.

Substances that produce hydroxide ions, OH^-, when dissolved in water are bases. Bases taste bitter, turn the color of litmus from red to blue and react with acids in aqueous solutions to produce salts. Bases dissolve non-metal oxides, forming anions containing oxygen.

Example 2.15. Identify the following compounds as acids, bases or salts.

a. CaOH **b.** H_3PO_4 **c.** $MgCl_2$ **d.** HCl

Solution:

a. a base, hydroxide ion will be produced;
b. an acid, hydrogen ions are produced;
c. a salt, a combination of a metal and a nonmetal to form an ionic compound;
d. an acid, hydrogen ions are released.

13. Naming simple acids and bases.

An acid produces H^+ and an anion. Acids are named based on the anion. We can name the acids based on the anion ending. The examples are given in the text.

Anion ending	Acid name
-ide	hydro______ic acid
hypo___ite	hypo_______ous acid
-ite	_________ous acid
-ate	_________ic acid
per____ate	per______ic acid

Example 2.16. Name the following acids.

a. HF **b.** H_2CO_3 **c.** H_2SO_4

a. The anion in this case is fluoride. The acid is therefore hydrofluoric acid.
b. The anion is carbonate. The -ate ending makes the acid carbonic acid.

c. The anion is sulfate. We change the name slightly for pronunciation, in this case -ur is added. The acid is sulfuric acid.

14. Naming simple organic compounds.

In this section we will limit our discussion to simple hydrocarbons, which contain only hydrogen and carbon. We will further limit ourselves to saturated hydrocarbons. These are compounds that contain the maximum number of hydrogen for each carbon atom. To name these simple compounds, we will start with linear compounds; that is, all the carbons are joined in a simple single chain with no branch points. These compounds are named for the number of carbon atoms using the prefixes for compounds used earlier and given in Table 2.6. The general formula for alkanes is $C_nH_{(2n+2}$.

Example 2.17. Name or write the formulas for the following compounds:

a. C_3H_8 **b.** pentane **c.** one of the C_8H_{18} isomers **d.** methane

Solution:

a. Three carbons indicate prop- for the prefix. The compound follows the general formula for alkanes and the -ane ending is used. The compound is propane.

b. C_5H_{12}. Using the pent- prefix, we know there are five carbon atoms. The number of hydrogen atoms is determined from the formula above.

c. Octane.

d. CH_4

Writing structural formulas for simple branched alkanes.

1. If an alkane has a branched carbon atom chain, count the number of carbon atoms in the longest continuous chain. This is the parent alkane and the compound will be named as a derivative of it.
2. Next, if the substituents (the groups that replace hydrogen as side chains) are also alkanes, then the substituents are named as alkane derivatives with the ending changed to -yl instead of -ane.
3. The location of the substituent is given by numbering the parent chain. The numbering is done so that the substituent will have the smallest number.

Example 2.18. Name the following compounds:

a. $CH_3CH_2CHCH_3$ (with CH_3 attached below the CH) **b.** $CH_3CH_2CH_2CHCH_3$ (with CH_2CH_3 attached below the CH)

Solution: For compound **a,** the longest carbon chain is four carbons. Therefore, the compound will end in butane. To name the substituent, the group has one carbon atom and is therefore methyl. (Note that in this case it doesn't matter if we count the longest chain as going across or as being three carbons across and the one carbon down.) The compound is methylbutane, but we still need to specify the position of the methyl. In order to give the lowest number, this would be 2-methylbutane, however the 2 is unnecessary for the following reasons. We would not have either 1- or 3- methylbutane. 1-methylbutane would actually be pentane and 3-methylbutane would not have the lowest number. However, if another group is present we might use the 3-methyl designation. For example, one could have 2,3-dimethylbutane.

For compound **b,** we proceed in the same way. The longest carbon chain is six so the compound will be hexane. (Don't be fooled by the way the formula is written.) The substituent is methyl and the location is the third carbon, numbering to give the lowest number. The compound is 3-methylhexane.

Example 2.19. Which of the following names for compounds are not possible and why? What would the correct name be?

a. 2-ethylpropane **b.** 3-methylbutane **c.** 1-methylethane

Solution: None of the names is correct. If you draw the structure for 2-ethylpropane, you will see that the longest carbon chain is actually four carbons and this should be named as methylbutane (see discussion above). 3-methylbutane is incorrect because the numbering to give the lowest number would be 2-methylbutane, which is unnecessary. In the third case, 1-methylethane, is actually propane. The carbon chain is continuous and is three carbons long.

Properties and structural formulas of some common organic functional groups.

Functional groups are atoms or groups of atoms attached to a hydrocarbon chain that confer characteristic properties. We will describe a few simple organic functional groups and some simple properties.

Alcohols: The functional group is the hydroxyl group, -OH. Compounds are named similar to the alkanes. The ending is changed to -ol instead of -ane. Therefore CH_3OH is methanol.

Carboxylic acids: The functional group is the carboxyl group (-COOH). Compounds are named similar to the alkanes. The ending is changed to -oic acid. Therefore, the two carbon acid, CH_3COOH is ethanoic acid.

Example 2.20. Name the following compounds

a. CH_3CH_2COOH **b.** CH_3CH_2OH **c.** CH_3NH_2

Solution:

a. There are three Cs, so the prefix is propyl; the functional group, COOH is an acid. The compound is propanoic acid. Naming is analogous to naming alkanes. The -yl of the prefix is changed to-an and the suffix is added for the functional group.

b. Ethanol. There are two C, and the OH is the alcohol functional group.

c. Methylamine. There is one C and an amine functional group.

Skills Test Problem

Hemoglobin, the oxygen carrier in red blood cells, contains 0.340% iron by mass. Each hemoglobin molecule contains four Fe atoms. Calculate the molar mass of hemoglobin.

Solution. First, iron has an atomic mass of 55.87. Four iron atoms make up 0.340% of the total mass of hemoglobin. Therefore

4 x 55.87 = (0.340/100) x Total mass of hemoglobin

Hemoglobin mass = (4 x 55.87)/(0.00340) = 65,729 = 65,700 rounding to sig. figures.

Note: Don't forget to divide by 100 just because the % is less than one.

Quiz A

1. A subatomic particle that has the same mass as the ^{1}H nucleus and a positive charge is called

   a. a proton
   b. a neutron
   c. an electron
   d. an isotope
   e. an alpha particle

2. An atom with the chemical symbol $^{35}_{17}X$ has

   a. 35 protons and 17 electrons
   b. 35 neutrons and 17 protons
   c. 17 protons and 18 neutrons
   d. 17 protons and 35 electrons
   e. none of the above

3. Name the following ionic compounds:

   a. KCN b. $CaCO_3$ c. Na_2SO_4

4. Give formulas for the following acids and bases:

a. hydrochloric acid
b. sulfuric acid
c. sodium hydroxide
d. phosphoric acid

5. When 4.0 g of methane is burned in 16.0 g of oxygen, 11.0 g of carbon dioxide and 9.0 g of water are produced. This illustrates

a. law of constant composition
b. the isotope effect
c. law of conservation of mass
d. acid base reaction
e. none of these

6. Isotopes of an element have
   a. the same number of protons
   b. the same number of neutrons
   c. more electrons than protons
   d. all of the above
   e. none of the above

7. Write the condensed structural formulas for the following compounds:
   a. 1-pentanol
   b. 2-methyl hexane
   c. butanoic acid

Quiz B

1. There are two stable isotopes of chlorine: Cl-35 = 34.9689 amu (75.53%) and Cl-37 = 39.9659 amu. What is the average atomic mass of chlorine?
   a. 34.9689 + 36.9659
   b. (34.9689 + 36.9659)/2
   c. (34.9689)(75.53) + (36.9659)(24.47)
   d. (34.9689)(75.53) + (36.9656)(24.47)}/100
   e. none of these

2. Two atoms with the same number of protons and a different number of neutrons are called:
   a. different elements
   b. ions
   c. radioactive
   d. isotopes
   e. none of these

3. Which of the following is **not** the correct name for an organic compound?
   a. ethanol
   b. 1-methylpropane
   c. 2-methylpropane
   d. 2-methylbutane
   e. all are incorrect

4. Name the following compounds:

a. NaOH

b. Li_2SO_4

c. HCl

d. CH_3OH

5. Write the formula for each of the following compounds:

   a. iron(III) chloride

   b. sulfur hexafluoride

   c. sulfuric acid

   d. sodium phosphate

6. In one compound 8.00 g of oxygen combine with 6.00 g of carbon and in a second compound 8.00 g of oxygen combine with 12.00 g of carbon. This is an example of:

   a. The law of definite composition

   b. The law of multiple proportions

   c. The law of conservation of mass

   d. The isotope effect

   e. None of these

7. An atom has an atomic mass of 35 and one more neutron than proton. Write the correct symbol for the atom.

Quiz C

1. The Law of Conservation of Mass

   a. states that two elements will always combine in the same ratio with each other

   b. was first stated by Dalton

   c. is a theory that remains to be proved

   d. states that the total mass remains constant during a chemical reaction

2. For carbon monoxide and carbon dioxide, the masses of oxygen that combine with a fixed mass of carbon are in small whole number ratios. This illustrates

   a. Dalton's Atomic Theory

   b. The Law of Conservation of Mass

   c. Chemical reactivity

   d. The law of definite proportions

   e. Acid-base reactions

3. The nucleus of an atom contains

   a. protons

   b. neutrons

   c. electrons

   d. protons, neutrons and electrons

   e. neutrons and electrons

f. protons and neutrons

4. Name the following compounds:
   a. $AlPO_3$
   b. CsI
   c. $Co_2(CO_3)_3$
   d. H_2S
5. The correct name for HCl (aq) is
   a. hydrogen chloride
   b. chlorite
   c. hypocholic acid
   d. hydrochloric acid
6. Write the formula for each of the following compounds:
   a. ethanol
   b. hexane
   c. propanol
   d. methylamine
   e. propanoic acid
7. The element silver consists of two isotopes: ^{107}Ag with an atomic mass of 106.905 and a natural abundance of 51.83%, and ^{109}Ag with an atomic mass of 108.905 amu and a natural abundance of 48.17%. Calculate the average atomic mass of silver.
8. Give the main property of a simple acid and a simple base.
9. A new element, hugeium, is discovered, which has two naturally occurring isotopes. One isotope has an atomic mass of 300 amu the other isotope has an atomic mass of 305 amu. The atomic mass of the element is 301. What general conclusion can you make about the relative abundance of the two isotopes?
10. List three ionic and three molecular compounds.

Self Test

1. Write the formula for each of the following compounds
   a. iron(III)periodate
   b. calcium fluoride
   c. ethanol
   d. hydrobromic acid
   e. potassium phosphite

Multiple Choice

2. The isotope $^{151}_{63}Eu$ has

a. 151 protons
b. 151 neutrons
c. 63 protons
d. 63 neutrons
e. none of the above

3. Dalton's Atomic Theory states that
   a. matter is conserved
   b. atoms of a given element are alike in mass and other properties
   c. a chemical reaction involves loss of protons and neutrons to form new elements
   d. isotopes of many elements exist

4. Elements to the left of the periodic table
   a. are classified as nonmetals
   b. have greater atomic mass
   c. have greater atomic number
   d. are classified as metal
   are unreactive

5. wo elements combine in different ways to form different substances, the mass os of the two elements in each are
   a. the same for the two compounds
   b. small, whole number multiples of each other
   c. the same number of protons, but different number of neutrons
   d. cannot be determined
   e. more accurate

True or False

6. Isotopes are atoms with identical mass numbers but different atomic numbers.
7. In a chemical reaction, the atoms are rearranged but are not broken apart or destroyed.
8. Elements differ from one another according to the number of electrons.
9. The isotopic mass represents the number of neutrons and protons in a particular atom.
10. The correct formula for bromous acid is HBrO.

Problems

11. A colorless liquid is thought to be a pure compound. Analysis of three samples of the material gives the following results:

Mass of Sample	Mass of Carbon	Mass of hydrogen
Sample 11.549 g	0.968 g	0.0649 g
Sample 20.244 g	0.229 g	0.0153 g

Sample 3 1.000 g	0.625 g	0.0419 g

Could the material be a pure compound? What law does this illustrate?

12. The atomic mass of Chlorine is 35.45 amu. Chlorine contains two isotopes. Assume the masses of Cl isotopes are integers. One isotope, ^{35}Cl has a natural abundance of 75.77%. What must the other isotope of Cl be?
13. A student heated 0.5000 g of zinc and 1.000 g of sulfur in a closed container. She obtains 0.308 g of zinc sulfide, ZnS. What mass of unreacted zinc must remain?
14. Lead consists of the following isotopes.

Isotope	Atomic mass, u	Percent Abundance
Lead-204	203.973	1.480
Lead-206	205.9745	23.60
Lead-207	206.9759	22.60
Lead-208	207.9766	52.30

Calculate the weighted-average atomic mass of lead.

15. Take the mass of an atom to be the sum of the masses of its protons, neutrons, and electrons. What is the approximate percentage by mass of the protons, neutrons, and electrons in ^{12}C?

Short Answer Questions:

16. Explain why a chemist calls the compound CaF_2 calcium fluoride and not calcium difluoride. Would a chemist call $FeCl_2$ iron chloride?
17. Write structural formulas for as many alkane isomers as you can having the formula C_5H_{12}.
18. How might you distinguish between an acid and a base in the laboratory?
19. Describe some of the properties of metals. Where are they located in the periodic table?

CHAPTER 3

Stoichiometry: Chemical Calculations

Stoichiometry problems focus on the relationships between reactants and products, how much will react, and how much if any will be left over. We have already seen that atoms are not created or destroyed in a chemical process. This concept is central to the idea of stoichiometry where the atoms are rearranged.

Chapter Objectives: By the time you have finished this chapter you should be able to:

1. Define and calculate molecular masses and formula masses.
2. Define a mole and be able to calculate and use molar mass.
3. Determine the mass percent composition from chemical formulas.
4. Determine the chemical formulas from mass percent composition.
5. Relate molecular formulas to empirical formulas.
6. Describe elemental analysis.
7. Write and balance chemical equations.
8. Use stoichiometric equivalents to determine limiting reagents in a reaction.
9. Define theoretical yield and percent yield.
10. Define molarity and calculate concentrations of solutions.
11. Calculate solution dilutions.

Chapter Summary

1.A. Defining and calculating molecular masses.

Remember that each atom has a characteristic atomic mass. This atomic mass is based on the atomic mass of each isotope and the abundance of each isotope as described in the previous chapter. The molecular mass is the mass of a molecule of a substance. Simply put, we add up the atomic masses of the elements composing the molecule. We must use total numbers of atoms. For example, CH_4 contains one carbon and four hydrogen atoms. The molecular mass is 1 x atomic mass of carbon + 4 x atomic mass of hydrogen.

Example 1. Calculate the molecular mass of the following compounds:

a. methane **b.** C_3H_7Cl

Solution:

a. First write the correct molecular formula. In this case, CH_4. This indicates one atom of carbon and four atoms of hydrogen. Next, look up the atomic mass of the elements. C = 12.011 u and H = 1.00794 u. We now add the atomic masses.

1 x atomic mass C + (4 x atomic mass H) = 12.011 u + (4 x 1.00794 u) = 16.043 u

b. This is calculated in a way similar to the above problem.

3 x atomic mass of C = 3 x 12.011 u = 36.033 u
7 x atomic mass of H = 7 x 1.00794 u = 7.05558 u

1 x atomic mass of Cl = 1 x 35.4527 u = 35.4527 u

Adding these gives a molecular weight of 78.54128 u and given the correct significant figures this is 78.541 u.

1.B. Defining and calculating formula masses.

The term molecular mass is appropriate only for compounds that exist as distinct molecules. For some substances, however, discrete molecules do not exist. These compounds are represented by a formula mass, which represents the simplest combination of atoms or ions consistent with the chemical formula. The formula mass will be the same as the molecular mass for those compounds that are composed of distinct molecules. The formula masses are calculated in the same way as molecular masses. We must again write the correct formula for the compound. We then look up the atomic mass of each of the atoms in the formula, and add the total number of atoms exactly like the above examples for molecular masses.

Example 2. Calculate the formula masses of each of the following compounds:

a. K_2SbF_5 **b.** barium bromide **c.** iron (III)phosphate **d.** $NaB(C_6H_5)_4$

Solution:

a. Adding up the atomic masses with the appropriate coefficients:

2 x atomic mass K + 1 x atomic mass Sb + 5 x atomic mass F =
(2 x 39.0983 u) + (121.75 u) + (5 x 18.9984 u) = 294.9386 u,
rounding to the correct significant figures gives 294.94 u.

b. The formula is $BaBr_2$

atomic mass Ba + (2 x atomic mass Br) = 137.327 u + (2 x 79.904 u) = 297.14 u

c. The formula is $FePO_4$

atomic mass Fe + atomic mass P + (4 x atomic mass O) =
55.847 u + 30.9738 u + (4 x 15.9994 u) = 150.8184 u = 150.82 u

d. In this case, the formula indicates 4 x C_6H_5. We can add (4 x 6 x atomic mass C) + (4 x 5 x atomic mass H) or we can add up the mass of C_6H_5 and then multiply by 4. We get the same answer, but remember there are 24 C, not 4 or 6. Calculating the formula mass of C_6H_5 gives (6 x atomic mass C) + (5 x atomic mass H) = (6 x 12.011 u) + (5 x 1.00794 u) = 77.106 u.

Now calculating the formula mass of $NaB(C_6H_5)_4$

atomic mass Na + atomic mass B + (4 x atomic mass C_6H_5) =
22.9898 u + 10.811 u + (4 x 77.106 u) = 342.22 u.

2.A. Definition of a mole.

A mole is defined as the amount of substance that contains as many elementary units as atoms in exactly 12 g of the carbon-12 isotope. A further definition is that a mole is 6.022137×10^{23} particles. This number is also known as Avogadro's number. Keep in mind that a mole is an amount in the same way that a dozen is twelve. A mole is Avogadro's number of an item. Since atoms and molecules are very small, this unit is convenient for measurement of these quantities.

2.B. Calculation and use of molar mass.

Since not all atoms or molecules have the same mass and a single atom or molecule is too small to weigh in the lab, it is convenient to use molar masses. The molar mass is the mass of one mole of a substance. If we use atomic or molecular weight, we can determine the molar mass. Simply change the unit u (of atomic, molecular or formula weight) to the unit g/mol to obtain the molar mass as in Example 3. Alternatively, sum the molar mass of each element multiplied by the number of moles of that element in one mole of compound, as in Example 2.

Example 3. Calculate the mass of 0.740 mol of KCl.

Solution: First calculate the formula weight of KCl. To do this we add the atomic weight of K + atomic weight of Cl. There are no coefficients to multiply since the formula unit contains only one atom of each element. The formula weight is:

$$39.0983\ \text{K} + 35.4527\ \text{Cl} = 74.5510\ \text{KCl}$$

This is also the number of grams/mole, the molar mass. Therefore, the mass of 0.74 mol of KCl is

$$0.740\ \text{mol KCl} \times \frac{74.551\ \text{g}}{1\ \text{mol KCl}} = 55.2\ \text{g KCl}$$

Example 4. Calculate the number of formula units in 0.650 g of NaOH.

Solution: First, we can calculate the number of moles from the molar mass. Next we can calculate the number of formula units from the number of formula units in a mole of substance.

The formula weight is 1 x mass of Na + 1 x mass of O + 1 x mass of H =
22.9898 u + 15.9994 u + 1.00794 u = 39.9971 u. The molar mass is 39.9971 g/mol.

To calculate the number of moles, use the molar mass as a conversion factor:

$$0.650\ \text{g of NaOH} \times \frac{1\ \text{mol NaOH}}{39.9971\ \text{g}} = 0.01625\ \text{mol}$$

To calculate the number of formula units, we use Avogadro's number:

$0.01625\ \text{mol} \times 6.022 \times 10^{23}$ formula units/mol $= 0.09786 \times 10^{23}$
using correct scientific notation, $= 9.79 \times 10^{21}$ formula units.

3. Mass percent composition from chemical formulas.

The mass percent is the fractional weight of each element in a compound multiplied by 100 to express the ratio as a percentage. To calculate this, take the total mass of each element, divide by the total mass of the compound. This gives the fractional weight of each element. To calculate the percent, multiply by 100.

Example 5. What is the mass percent of oxygen in H_2SO_4?

Solution: Calculate the mass of oxygen in the compound:

$$4 \text{ x mass of O} = 4 \text{ x } 15.9994 = 63.9976 \text{ g O/mol } H_2SO_4$$

The mass of H_2SO_4 is:

$$2 \text{ x mass H} + \text{mass of S} + 4 \text{ x mass of O} =$$

$$2 \text{ x } 1.00794 + 32.066 + 4 \text{ x } 15.9994 = 98.079 \text{ g/mol } H_2SO_4$$

The mass percent of oxygen is:

$$\frac{63.9976 \text{ g O}}{\text{mol } H_2SO_4} \text{ x } \frac{\text{mol } H_2SO_4}{98.079 \text{ g}} \text{ x } 100\,\% = 65.251\,\%$$

Note that the mass percent of an element in a compound must be less than 100. If you used the ratio incorrectly and inverted the answer, you would have a percentage of oxygen larger than the total percent of 100 %.

4. Chemical formulas from mass percent composition.

In order to relate mass percent to the chemical formula, we must first convert to a mole basis. Formulas are based on the relative number of moles of each element. Using the percent of each element, calculate the mass of each element based on a 100 g sample. Use the atomic weight to determine the number of moles. This will give the empirical formula. Since different compounds can have the same % composition and the same empirical formula, we need to know the molecular weight in order to determine the molecular formula.

Example 6. A compound contains 35.00% N, 5.05% H and 59.96% O. What is the formula for the compound?

Solution: First assume a 100.00 g sample and calculate the mass of each element. This gives 35.00 g N, 5.05 g H and 59.96 g O. Next find the number of moles of each element:

$$35.00 \text{ g x } \frac{1 \text{ mole}}{14.01\text{g}} = 2.498 \text{ mole N}$$

$$5.05 \text{ g x } \frac{1 \text{ mole}}{1.01\text{g}} = 5.00 \text{ mole H}$$

$$59.96\ g \times \frac{1 mole}{16.00 g} = 3.748\ mole\ O$$

We cannot have a formula with fractional subscripts, so we must somehow convert to whole numbers. The easiest way to do this is first divide by the smallest number, in this case 2.498. Doing this we get 1.00 mole N, 2.00 mole of H and 1.500, which rounds to 1.50, mole O. We now proceed as we did earlier and multiply by 2 to give 2 moles of N, 4 moles of H and 3 moles of O. The formula is $N_2H_4O_3$, or more likely NH_4NO_3. Often the formula coefficients do not come out to be exactly whole numbers so we need to know when we can round off. This will depend on how precisely the elemental analysis was done. In general, we will round one or two hundredths (2.98 becomes 3.00).

5. Relating molecular formulas to empirical formulas.

Molecular formulas are either the same as the empirical formula or multiples of them. If we have the molecular mass of a substance, we can determine the molecular formula from the empirical formula.

Example 7. A substance has the empirical formula CH_2. The molecular mass of the substance is 56.0 u. What is the molecular formula?

Solution: First calculate the mass of the empirical formula. In this case the mass is 14.0 u. Next divide the molecular mass by the empirical mass. The result is 4. Last, multiply the empirical formula by the factor to get the molecular formula. The molecular formula is C_4H_8.

6. Describing elemental analysis and determining the percentage composition of a compound from experimental data.

The composition of compounds containing C, H and O can be determined by burning a known mass of the compounds and determining the mass of products produced. The general reaction is :

$$C_xH_yO_z + w\ O_2 \text{-->} x\ CO_2 + y/2\ H_2O$$

Note that each mole of carbon in the unknown sample produces one mole of carbon dioxide and hydrogen goes only into water. Each mole of water contains two moles of H so that 1/2 mole of water is produced for each mole of H in the unknown. Convert the mass of products to moles and use this to determine the moles and mass of the C or H in the unknown. Since oxygen is in both products as well as available for combustion, the oxygen in the compound is found by subtracting the mass of hydrogen and carbon from the total mass. We do not need to determine the coefficient of O_2, w, except to balance the equation.

Example 8. A 0.2000-g sample of ibuprofen is burned to yield 0.5547 g of carbon dioxide and 0.1573 g of water. (a) Calculate the mass percent composition of ibuprofen. (b) Calculate the empirical formula.

Solution: We know that all the carbon in ibuprofen ends up as carbon dioxide, therefore, we calculate the number of moles of CO_2 and this will be the number of moles of carbon in the unknown sample.

$$0.5547 \text{ g x } \frac{1 \text{ mol } CO_2}{44.010 \text{ g } CO_2} = 0.01260 \text{ mol } CO_2$$

This is also the number of moles of C present. We now convert this to the mass of carbon:

$$0.01260 \text{ mol C x } \frac{12.011 \text{g C}}{1 \text{ mol C}} = 0.1514 \text{ g C}$$

The mass percent of carbon in the sample can be calculated from the mass of carbon and the mass of the sample.

$$\% \text{ C} = \frac{0.1514 \text{ g C}}{0.200 \text{ g sample}} \text{ x } 100 = 75.70 \text{ \% C.}$$

Repeat these steps to find the mass percent of H.

$$0.1573 \text{ g x } \frac{1 \text{ mol } H_2O}{18.016 \text{g } H_2O} = 0.008732 \text{ mol } H_2O$$

Since each mole of water contains two moles of H, the number of moles of hydrogen in the ibuprofen sample is twice that in the H_2O.

$0.008732 \text{ x } 2 = 0.01746$ mol H in ibuprofen.

To calculate the mass:

$$0.01746 \text{ mol x } \frac{1.0079 \text{g H}}{1 \text{ mol H}} = 0.01760 \text{ g H}$$ and calculating the percent H we get

$$\frac{0.01760 \text{ g H}}{0.200 \text{ g sample}} \text{ x } 100 \text{ \%} = 8.80 \text{ \% H.}$$

The last step is to calculate the percent oxygen from the difference.

100% - 75.70% C - 8.80 % H = 15.50 % Oxygen

To find the empirical formula of the compound we proceed as we did earlier. First find the moles of oxygen

$$\frac{15.5 \text{ \% O}}{100 \text{ \%}} \text{ x } 0.200 \text{ g x } \frac{1 \text{ mol O}}{15.9994 \text{ g}} = 0.001938 \text{ mol O}$$

and then divide the moles of each element by the smallest number and convert to whole number ratios.

$C_{0.01260}H_{0.01746}O_{0.001938}$ dividing by 0.001938 gives

$C_{6.5}H_9O$ and multiplying by 2

$C_{13}H_{18}O_2$ for the formula.

7. Writing and balancing chemical equations.

A chemical equation is a way of representing chemical change. The reactants are written on the left and are the starting material. The products of the reaction are written on the right and are the final material. We use chemical symbols to represent the compounds and elements in the reaction. We use numbers in front of the symbols to show the number of molecules or atoms of each substance that is involved in the reaction. Because of conservation of matter, we know that the number of atoms of each element on the left side of the equation must be the same as the number of atoms on the right. These atoms will of course recombine into different compounds, but the number of atoms of each element will be the same. For example, the reaction given below:

$$CH_4 + O_2 \text{-->} CO_2 + H_2O$$

This equation is not balanced since there are two atoms of oxygen and four atoms of hydrogen on the left and on the right there are three atoms of oxygen and two atoms of hydrogen. The number of atoms of carbon is the same on both sides of the reaction. In order to balance the equation, we must add coefficients to some of the molecules. Generally, it is easier to leave the atoms that do not occur in compounds (oxygen in this case) last. Since carbon is already balanced, we balance the hydrogen by multiplying H_2O by 2. We now have four oxygen on the right and so multiply O_2 by two. The final equation is then:

$$CH_4 + 2\ O_2 \text{-->} CO_2 + 2\ H_2O$$

Remember when balancing equations that the number of atoms is determined by both the coefficient and the subscript (2 H_2O has four atoms of H). It is important to remember that you balance equations by changing the coefficients, NOT the formula(s) of any compound.

Example 9. Balance the following chemical equation:

$$Pb(s) + PbO_2(s) + H_2SO_4(l) \rightarrow PbSO_4\,(s) + H_2O$$

Solution: We can start by balancing Pb first. We have 2 Pb on the left as reactants and one as product. Using a coefficient of 2 for $PbSO_4$ (s) balances Pb. We now have

$$Pb(s) + PbO_2(s) + H_2SO_4(l) \rightarrow 2PbSO_4\,(s) + H_2O$$

Looking next at S, we have two on the left and only one on the right side of the equation. Therefore we use the coefficient 2 in front of $H_2SO_4(l)$. This now gives

$$Pb(s) + PbO_2(s) + 2\ H_2SO_4(l) \rightarrow 2PbSO_4\,(s) + H_2O$$

We are left to balance the oxygen and hydrogen. We have four H on the left (there are two H in each H_2SO_4 and we have two molecules, therefore, four H) and two on the right. Using a coefficient of 2 for H_2O, we balance the H. If we now check the O, we have 10 on the left and ten on the right. The balanced equation is

$$Pb(s) + PbO_2(s) + 2\ H_2SO_4(l) \rightarrow 2PbSO_4\,(s) + 2H_2O$$

8.A. Use chemical equations to determine the stoichiometry of reactions.

As we have seen in calculating how many moles of CO_2 are produced by burning, the chemical equations give us the relationships between the moles of reactants and moles of products. Keep in mind that the equations must be balanced to use this relationship and that the chemical equation relates the number of moles.

Example 10. Calculate the number of moles of O_2 that react with 7.32 moles of C_3H_8O when the sample is burned.

Solution: We are given the amount in moles of C_3H_8O. We need to relate the moles of oxygen that will react and to do this we write a balanced chemical equation.

From previous problems we know that the products of combustion of a compound that contains C, H and O are CO_2 and H_2O. The reaction is:

$$C_3H_8O + O_2 \text{-->} CO_2 + H_2O$$

We need to balance the equation so that the number of moles of each element in the product is equal to the number of moles in the reactant. That is, the number of moles of each element on the left is equal to the number of moles of the same element on the right. Start by balancing the C :

$$C_3H_8O + O_2 \text{-->} \mathbf{3}\ CO_2 + H_2O$$

Next balance the H:

$$C_3H_8O + O_2 \text{-->} 3\ CO_2 + \mathbf{4}\ H_2O$$

Finally balance the O by using a coefficient for O_2 (for diatomic gases we can use 1/2 as a coefficient):

$$C_3H_8O + 4\ 1/2\ O_2 \text{-->} 3\ CO_2 + 4\ H_2O$$

It would also be correct to balance with whole numbers and we would get:

$$2\ C_3H_8O + 9\ O_2 \text{-->} 6\ CO_2 + 8\ H_2O$$

We now have a relationship between the number of moles of oxygen that react with each mole of C_3H_8O, 9:2 or 4.5:1.

Therefore, the number of moles of oxygen that are required are:

$$7.32 \text{ moles of } C_3H_8O \times \frac{9 \text{ moles } O_2}{2 \text{ moles } C_3H_8O} = 32.9 \text{ moles } O_2$$

We would get the same result if we multiplied by 4.5 and divided by 1 for the ratios from the first equation.

8.B. Limiting reagents in a chemical reaction.

The limiting reagent is the substance we run out of first in a chemical reaction and therefore limits the amount of product that can be formed. If you are baking cookies and have a quart of milk and a dozen eggs, you will probably run out of one before the other is completely used up. This is the limiting reagent. In a more quantitative way for a chemical reaction, we can determine which reagent is the limiting reagent. If we think about our cookie example, suppose each dozen cookies requires 4 eggs and 1 cup of milk. From our dozen eggs, we can make 3 dozen cookies. But do we have enough milk? Each dozen requires one cup of milk and one quart contains 2 pints and each pint contains 2 cups or we have a total of 4 cups. This would make four dozen cookies, but we only have enough eggs for three dozen. The eggs are the limiting reagent! What we have just calculated uses the same technique we use for chemical reactions. First we need to write a balanced chemical equation to relate the amount of the two reactants. We then calculate how much of the other reactants we would need to completely use up the reactant in question. Is there enough? If not, which reactant(s) are we short? If we have enough of all but one reactant, this is the limiting reagent. If two reactants are lacking, then we need to calculate again and determine which we would run out of first.

Let's assume our cookie recipe also requires 1/2 lb of butter for each dozen cookies and we have one pound of butter. We already know that we have enough milk and we have eggs for three dozen cookies. However, we only have enough butter for two dozen so in this case butter is the limiting reagent. (Don't try this recipe at home; a few other ingredients are required!)

Exercise (text) 3.20A. Hydrogen sulfide is produced by the reaction

$$FeS(s) + 2\ HCl(aq) \rightarrow FeCl_2(aq) + H_2S$$

If 10.2 g HCl is added to 13.2 g FeS, how many grams of H_2S can be formed?
Solution: First check if the equation is balanced, which it is. Next convert grams to moles, since equations are based on moles of reactants and products.

$$13.2\text{ g FeS} \times \frac{1\text{ mol FeS}}{87.913\text{ g}} = 0.150\text{ mol FeS}$$

$$10.2\text{ g HCl} \times \frac{1\text{ mol HCl}}{36.4606\text{ g}} = 0.280\text{ mol HCl}$$

From the reaction stoichiometry, to use up all of the FeS, we need two moles HCl: 1 mole FeS. Therefore,

$$1.50\text{ mol FeS} \times \frac{2\text{ mol HCl}}{1\text{ mol FeS}} = 0.300\text{ mol HCl}$$

However, we only have 0.280 mol HCl. We cannot use up all the FeS because HCl is limiting. We can now calculate, using the reaction stoichiometry, the amount of product obtained by using all the HCl:

$$0.280\text{ mol HCl} \times \frac{1\text{ mol }H_2S}{2\text{ mol HCl}} \times \frac{34.082\text{ g }H_2S}{1\text{ mol }H_2S} = 4.77\text{ g }H_2S$$

9. Define theoretical yield and percent yield. Calculate the percent yield from the theoretical and actual yield.

The theoretical yield is the amount of product that could be obtained from the amount of reagents if the reaction went to completion. It is the maximum amount of product that could be obtained. The actual yield (what you actually produced) can be less than the theoretical yield for a variety of reasons described in the text. The percent yield is the actual yield divided by the theoretical yield multiplied by 100 to give percent.

10. Define and use the term molarity. Define dilute solutions and concentrated solutions.

Molarity is the number of moles of solute/liters of solution. Note that this is the number of liters of solution, not just the solvent. For many solutions we will be given the weight of solute and the volume of solvent. In some cases, we can ignore the volume of the solute, but this is not always the case.

Example 12. Calculate the molarity of the following solution: 18.0 mol H_2SO_4 in 2.00 L of solution.

Solution: We already know the number of moles so we need only divide the number of moles by the volume of solution. This will give a 9.00 molar solution.

Example 13. A 5.00 g sample of NaCl is dissolved in water to give a total volume of solution of 1.00 L. What is the molarity of the solution?

Solution: We need to know the volume of solution and the moles of solute (NaCl). We are already given the volume of solution. We calculate the moles of solute by dividing the number of grams of NaCl by the molar mass. The molar mass of NaCl is 22.99 u + 35.45 u = 58.44 g/mol. This is the sum of the atomic weights of Na and Cl taken from the periodic table. The number of moles is

$$5.00 \text{ g NaCl} \times \frac{1 \text{ mole NaCl}}{58.44 \text{ g}} = 0.0856 \text{ mole.}$$

Now divide the number of moles by the volume of solution, 1.00 L. This gives 0.0856 moles/L or 0.0856 M.

11. Calculate solution dilutions.

A simple relationship exists for calculating solution dilutions. The moles of solute do not change as we are diluting with solvent. (In the case where we mix two solutions of different molarity the calculation is more complex, but we will only consider dilution with solvent for now.) The total number of moles of solute is the same before and after dilution. The equation for the moles of solute is that moles of solute = molarity of solution x volume. Therefore the M of the concentrated solution times the volume will be equal to the M of the diluted solution times the volume of the dilute solution.

$$M_{conc} \times V_{conc} = M_{dil} \times V_{dil}$$

This will give the same result as the method described on p. 98 of the text. We have simply combined the algebra.

Example 14. What volume of the above 0.0856 M NaCl solution would be required to make 1.00 L of 0.015 M NaCl?

Solution: Using the above relationship, we know M_{conc}, M_{dil} and V_{dil}. We need only to find the volume of the concentrated solution to use. Dividing both sides by M_{conc} we get:

$$V_{conc} = \frac{M_{dil} \times V_{dil}}{M_{conc}} = \frac{0.015\text{M} \times 1.00\text{L}}{0.0856\text{M}} = 0.175\text{ L}$$

Note that the volume is less than one liter and about 20% of the final volume which a common sense estimate would suggest is about right. If you had carried out the alegbra incorrectly, you might have gotten an answer larger than one liter which of course is not correct. Don't forget to estimate your answers and check if they make sense.

Exercise (text) 3.25A. A stock bottle of aqueous formic acid indicates that the solution is 90.0% HCOOH by mass and has a density of 1.20 g/mL. Calculate the molarity of the solution.

Solution: This is a multi-step problem that involves conversion factors. We know the density and the mass percent of the solution. We need to convert to moles/liter. If we look at the density first, we have g/mL. This relates the mass of the solution to the volume. We can first convert this to g/L by multiplying by 1000 mL/L to give 1200 g/L. Now we need to know how many moles this contains. Only 90.0% of the solution is solute so 90.0 g solute/100 g of solution. We have now related the mass of solute to mass of solution and from the density we can relate mass of solution to volume. We can then get mass of solute/L of solution and convert to moles of solute per liter of solution. This outline tells us how to go about solving the problem. The solution is worked out below:

$$\frac{90\text{ g solute}}{100\text{ g solution}} \times \frac{1.20\text{g solution}}{\text{mL}} \times \frac{1000\text{mL}}{\text{L}} \times \frac{1\text{mole HCOOH}}{46.0257\text{g}} = 23.5\text{ moles/Liter or } 23.5\text{ M.}$$

Skills Test Problem

3.1. A sample of 15.0 g of sugar (molar mass 342 g/mol) is dissolved in 100 g of water. The resulting solution was found to have a density of 1.08 g/cm^3. What is the molarity of the solution?

Answer: First, the definition of molarity is number of moles/liter solution. To calculate the volume of solution, use the density.

Mass of solution = 15.0 g sugar + 100 g H_2O = 115 g

$$\text{Volume} = 115\text{g} \times \frac{1\text{ cm}^3}{1.08\text{g}} \times \frac{1\text{ L}}{10^3\text{ cm}^3} = 0.106\text{ L}$$

$$\text{moles of solute} = 15.0 \text{ g x } \frac{1\text{ mole}}{342\text{ g}} = 0.0438 \text{ g}$$

$$\text{molarity} = \frac{0.0438\text{ moles}}{0.106\text{ L}} = \mathbf{0.414\ M}$$

Quiz A

1. What is the percent composition of $C_6H_{12}O_6$?
2. A solution containing 4.83 g of NaCl is mixed with a solution containing 6.78 g of $AgNO_3$. A precipitate of AgCl forms. Assuming that none of the AgCl remains in solution, how many grams of AgCl form?
3. What is the molar mass of NH_4Cl?
4. A 25.0 g sample of a compound contains 11.85 g of carbon, 2.625 g of hydrogen, and 10.525 g of oxygen. What is the simplest formula for the compound?
5. Balance the equation below:

   $NH_3 + O_2$ --> $NO + H_2O$

6. What mass of O_2, in grams, is required to completely burn 1.89 g of C_2H_5OH?
7. What is the molarity of a solution that contains 0.076 mol HCl in 1200 mL of solution?
8. Calculate the theoretical yield when 0.200 g Al reacts with 0.500 g O_2 according to the equation:

   $2\ Al + 3/2\ O_2$ ---> Al_2O_3

9. What mass of ammonia, in grams, can be made from 150 g H_2?

   $N_2 + H_2$ --> NH_3 (not balanced).

Quiz B

1. What is the molar mass of $Hg(NO_3)_2$?
2. A compound contains 63.6% N and 36.4% O. What is its empirical formula?
3. How many molecules are contained in 5.00 g of H_2O?
4. What volume of 0.379 M NaCl is needed to make 1.36 L of 0.231 M NaCl?
5. Write a balanced equation for the following reaction:

   Aluminum chloride + sodium hydroxide --> aluminum hydroxide + sodium chloride

6. 12.3 g of P_4 reacts with Cl_2 to produce 50.0 g of PCl_3. What is the percent yield?
7. Calculate the molarity of a solution of 3.00 mol NaCl in 800 mL of solution.
8. Balance the following equation:

$BF_3 + H_2O \rightarrow H_3BO_3 + HBF_4$

9. Calculate the mass in grams of 0.00200 mol CaH_2.

Quiz C

True or False

1. When hydrocarbons undergo combustion, carbon dioxide and water arc produced.
2. The empirical formula and the molecular formula of a compound are the same.
3. The limiting reactant determines the extent to which a chemical reaction takes place.
4. Elemental analysis can be used to determine whether a solution of HCl is a dilute solution or a concentrated solution.
5. The actual yield of a chemical reaction is always larger than the theoretical yield.

Questions

6. Write a balanced equation for the combustion of ethane.
7. What is the formula weight of calcium fluoride?
8. What is the molarity of a solution prepared by dissolving 1.45 g of NaCl in 2.40 L of solution?
9. A compound was found to contain 23 g of Na, 1.0 g H, 12 g C and 48 g of O. What is the empirical formula for the compound? Can you tell if this is the molecular formula?
10. How many moles are present in 9.0 g of $Ba_3(PO_4)_2$?

Self Test

Matching

1. mole
2. molarity
3. theoretical yield
4. chemical equation
5. empirical formula
6. combustion reaction
7. Avogadro's number
8. microscopic level

a. the number of moles of solute dissolved in one liter of solution
b. reaction with oxygen
c. gives the ratio of atoms in a compound
d. the mass of one mole of a substance
e. the amount of substance that contains as many elementary units as there are atoms in twelve g of carbon 12.
f. The number of elementary units in a mole, 6.022×10^{23}.
g. A description of a chemical reaction that uses symbols and formulas to represent the elements and compounds involved in the reaction.
h. The calculated quantity of a product expected in a reaction

9. solvent — i. The solution component in which one or more solutes are dissolved.

10. molar mass — j. The behavior of individual atoms and molecules represented by chemical symbols

Problems

11. In the following reaction:

    $PCl_3 + H_2O \rightarrow H_3PO_3 + HCl$ (not balanced)

    How many grams of H_3PO_3 can be formed from 0.89 g of PCl_3, if it is the limiting reagent? Assume the % yield was 65%. How many grams of H_3PO_4 were isolated?

12. How many grams of NaOH are required to make 250 mL of an 0.5 M solution?

13. Balance the following equation

    $AgNO_3 + Na_2SO_4 \rightarrow Ag_2SO_4 + NaNO_3$

14. Elemental analysis of a compound reveals its percent composition to be 50.7% C, 4.25% H and 45.1% O. Determine the empirical formula. The molecular weight of the compound was determined to be 141 g/mole. What is the molecular formula for the compound?

15. What volume of 12 M HCl is required to make 100 mL of a 0.25 M HCl solution?

16. Calculate the mass in g of 3.45 mol of NaOH.

17. Estimate which of the following has the greatest mass: 0.85 mol CH_3OH; 7.1 x 10^{23} molecules of CH_3OH; the mass of CH_3OH contained in 1.00 L of 1.5M CH_3OH.

18. In the following reaction

    $CaCN_2(s) + H_2O(l) \rightarrow CaCO_3(s) + NH_3$ (l) (not balanced)

    How many moles of $CaCO_3(s)$ will be produced from 3.2 g of $CaCN_2(s)$ and 2.5 g of H_2O ?

19. A woman wishes to take 200 mg of calcium as a dietary supplement. How many mg of a common antacid which is 75% calcium carbonate should she take?

20. What is the mass percent of C and H in C_4H_8 ?

CHAPTER 4

Chemical Reactions in Aqueous Solutions

Aqueous reactions, those carried out in water, have many important applications both in nature and in the laboratory. This chapter will explore a few basic categories of aqueous reactions and their applications.

Chapter Objectives: After completing this chapter you should be able to

1. Define an electroyte, anode, cathode, anion and cation
2. Calculate ion concentrations in solution
3. Know the difference between strong and weak acids and strong and weak bases
4. Define polyprotic acids and give an example of a polyprotic acid
5. Describe neutralization reactions
6. Predict whether or not precipitation will occur in a reaction. Know the solubility rules in Table 4.3.
7. Know the rules for assigning oxidation states
8. Know the rules for balancing oxidation-reduction reactions
9. Define disproportionation reactions
10. Define oxidizing and reducing agents
11. Describe some practical applications of oxidation and reduction reactions

Chapter Summary:

1.a. Definiton of an electrolyte.

Solutes of aqueous solutions can be classified according to their ability to conduct electrical current. If electrical conductivity is high, the solution is a strong electrolyte. Water soluble ionic compounds are strong electrolytes. Most molecular compounds are either nonelectrolytes or weak electrolytes. A strong electrolyte dissociates completely into ions.

1.b. Electrodes conduct electricity when partially immersed in electrolyte solutions and connected to an electrical source. The **anode** is the electrode connected to the positive pole. The **cathode** is connected to the negative pole.

1.c. An **anion** is a species carrying a negative (-) charge and is attracted to the anode. Examples: Cl^-, OH^-, NO_3^-.
A **cation** is a species carrying a positive (+) charge and is attracted to the cathode. Examples: H^+, Na^+, Fe^{2+}.

Example 4.1. A solution contains 3.5g of NaCl, 1.5g of KCl, 2.9g, of sodium citrate ($Na_3C_6H_5O_7$) and 20.0 g of glucose ($C_6H_{12}O_6$) per liter. Calculate the molarity of each of the species present in the solution.

Solution: We can calculate the molarity of each species by calculating the molarity from each source and adding them together. For example, Na^+ comes from both NaCl and sodium citrate. Calculating the molarity of Na^+ and Cl^- from NaCl gives:

3.5 g x 1 mol/58.44 g = 0.060 mol and there is one mole of Na^+/mol NaCl. The same is true for Cl^-. To get the total Na^+ concentration we need to include the Na^+ from sodium citrate. 2.9 g x 1 mol/258 g x 3mol Na^+/1 mol sodium citrate = 0.034 mol Na^+. The molarity of Na^+ is 0.060 mol/L + 0.034 mol/L = 0.094 M.

Citrate ion concentration from sodium citrate is one third that of Na^+ or 0.011 M. To calculate the Cl^- concentration from KCl, 1.5 g x 1 mol/74.6 g = 0.020 M and there is one mol K^+ for each mol of KCl. The Cl^- concentration is 0.020 M from KCl and 0.060 M from NaCl. The total $[Cl^-]$ = 0.020 M + 0.060 M = 0.080 M. The glucose concentration is 20.0 g x 1 mol/180 g = 0.11 M. Glucose does not ionize.

2. Differences between strong and weak acids and strong and weak bases.

Acids that are completely ionized in water solution are strong electrolytes and are strong acids, such as HCl. Most acids are partially ionized and are weak acids. Similarly, strong bases are completely ionized to produce hydroxide ions and weak bases are only partially ionized. Examples of weak bases are ammonia and most organic amines.

Example 4.2. State whether each of the following is an acid or a base and whether it is strong or weak, respectively.

a. NaCl **b.** HCl **c.** H_2SO_4 **d.** CH_3NH_2 **e.** NaOH

Solution:

a. NaCl is not an acid or a base; it is a strong electrolyte.

b. HCl dissociates completely and is a strong acid. It contains a Group 7A hydride, which are strong acids.

c. H_2SO_4 is a strong acid in its first ionization, but a weak acid in the second ionization.

d. Methylamine is a weak base as are most organic amines.

e. NaOH is a strong base. It is a group 1A hydroxide.

3. Definition of polyprotic acids.

Polyprotic acids can produce more than one hydrogen ion per molecule. These may be strong acids in the first ionization and weak acids in the second ionization. Examples are H_2SO_4 and H_3PO_4.

4.A. Neutralization reactions.

The reaction of an acid and base is a neutralization reaction. The acid and base react to produce water and a salt. If no other ions in solution react, the reaction can be written as:

$H^+ + OH^- \rightarrow H_2O$

Example 4.3. Calcium hydroxide is used to neutralize a waste stream of hydrochloric acid. Write **(a)** "complete formula" **(b)** ionic, and **(c)** net ionic equations for the neutralization reaction.

Solution:

a. The complete formula equation can be written:

$Ca(OH)_2 + 2\ HCl \dashrightarrow 2\ H_2O + CaCl_2$

b. The ionic equation is written:

$Ca^{2+} + 2\ OH^- + 2\ H^+ + 2\ Cl^- \dashrightarrow 2\ H_2O + Ca^{2+} + 2\ Cl^-$

c. The net ionic equation will cancel the spectator ions, Ca^{2+} and Cl^-, leaving:

$2\ H^+ + 2\ OH^- \dashrightarrow 2\ H_2O$

Cancel the 2 to give the simplest equation:

$H^+ + OH^- \dashrightarrow H_2O$

4.B. Titrations.

A technique in which two reactants in solution combine in stoichiometric proportions is a titration. In the case of an acid and a base, this means that the amount of acid will exactly neutralize the base so that for each hydroxide ion, one hydrogen ion is added.

Example 4.4. What volume of 0.550 M NaOH is required to titrate a 10.00 mL sample of vinegar that is 4.12% by mass of acetic acid, $HC_2H_3O_2$? Assume that the vinegar has a density of 1.01 g/mL.

Solution: The net ionic equation is:

$HC_2H_3O_2 + OH^- \dashrightarrow H_2O + C_2H_3O_2^-$

Calculate how many moles of acetic acid are present. Next, calculate how many moles of NaOH are required. From the number of moles required, calculate the volume required. The moles of acetic acid are:

$$10.00\ \text{mL} \times \frac{1.01\ \text{g solution}}{\text{mL}} \times \frac{0.0412\ \text{g acetic acid}}{1\ \text{g solution}} \times \frac{1\ \text{mol acetic acid}}{60.0\ \text{g}}$$

$= 0.00694$ mol acetic acid

Each mole of acetic acid requires one mole of NaOH for neutralization; therefore, 6.94×10^{-3} mol NaOH are required. To find the volume:

6.94×10^{-3} mol NaOH x 1 L/0.550 mol = 0.0126 L = 12.6 mL NaOH

5. Predicting whether or not precipitation will occur in a reaction.

In order to predict whether or not a precipitate will form in a reaction, we need to know the solubility rules. Using these rules, we can determine precipitation. The procedure is as follows: First write the possible ions in the reactant side of the equation. Second, look at possible combinations of positive and negative ions in the product side. Third, determine whether any of the combinations are insoluble according to the solubility rules. If any combination is insoluble, then a precipitate will form.

Example 4.5. Predict whether a reaction will occur in each of the following cases. If so, write a net ionic equation for the reaction:

a. $MgSO_4 + KOH$ -->

b. $FeCl_3 + Na_2S$ -->

c. $CaCO_3 + NaCl$ -->

Solution: In each case, we need to determine the possible combination of ions and which if any of the resulting compounds are insoluble. We can write all of the ions present from the two compounds. We then look to determine which possible anions and cations can combine. If the resulting combination of anion and cation yields an insoluble compound, a precipitate will form.

a. The possible products are $Mg(OH)_2$ and K_2SO_4. Using the solubility rules, we see that $Mg(OH)_2$ is insoluble. Potassium sulfate is soluble. The product is therefore $Mg(OH)_2$ and the reaction is:

$Mg^{2+} + SO_4^{2-} + 2K^+ + 2OH^-$ --> $Mg(OH)_2 + 2K^+ + SO_4^{2-}$

and the net ionic equation is:

$Mg^{2+} + 2OH^-$ --> $Mg(OH)_2$.

b. $2\ Fe^{3+} + 6\ Cl^- + 6\ Na^+ + 3\ S^{2-}$ --> $Fe_2S_3 + 6\ Cl^- + 6\ Na^+$

and the net ionic equation is:

$2\ Fe^{+3} + 3\ S^{2-}$ --> Fe_2S_3.

c. $CaCO_3$ + NaCl --> No reaction will occur since $CaCO_3$ is insoluble. Sodium carbonate is soluble.

6. Rules for assigning oxidation states.

Oxidation state or oxidation number refers to the number of electrons transferred, shared, or otherwise involved in the formation of the chemical bonds of a substance. The total of the oxidation states of atoms in a neutral species is zero. The most electronegative atoms should have negative oxidation states. Review the rules for oxidation numbers given in the text.

Example 4.6. What is the oxidation state of each element in the following?

a. Al_2O_3 **b.** P_4 **c.** $NaMnO_4$ **d.** ClO^- **e.** $H_2AsO_4^-$

f. $HSbF_6$ **g.** CsO_2 **h.** CH_3F **i.** $CHCl_3$

Solution:

a. Oxygen has an oxidation state of –2 (rule 5). In order for the compound to be neutral, Al has an oxidation state of +3.

b. For the molecule to be neutral, the oxidation state must be zero (rule 1).

c. From the rules for assigning oxidation numbers, Na has an oxidation state of +1 (rule 2). Oxygen has an oxidation state of –2 (rule 5). This gives -8 + 1 = -7 and therefore, Mn has an oxidation state of +7.

d. O has an oxidation state of –2 (rule 5). Chlorine must have an oxidation state of +1 since the ion has a charge of -1.

e. Hydrogen has an oxidation state of +1(rule 4) and oxygen of –2 (rule 5). As has an oxidation state of +5.

f. has an oxidation state of –1(rule 3) for a total of -6. H has an oxidation state of +1(rule 4). This means that in order for the molecule to be neutral, Sb has an oxidation state of +5.

g. Cs is a group 1A metal and has a +1 oxidation state (rule 2). O therefore has an oxidation state of -1/2.

h. F has a oxidation state of –1(rule 5) and H +1(rule 4). C must have an oxidation state of -2.

i. Cl has an oxidation state of -1 and H +1. C is +2.

7.A. Defining oxidation and reduction.

Oxidation involves an increase in the oxidation state of one or more elements. Reduction involves a decrease in the oxidation state of one or more atoms. Oxidation-reduction must occur together.

Example 4.7. Is the chemical reaction represented by the following equation a plausible one?

$$I_2(s) + Cl_2(g) + H_2O \text{ --> } IO_3^-(aq) + H^+(aq) + Cl^-(aq)$$

Solution: To determine whether the reaction is plausible, we must determine whether both oxidation and reduction are taking place. To do so, we write the oxidation numbers for each species:

$$I_2(s) + Cl_2(g) + H_2O \text{ --> } IO_3^-(aq) + H^+(aq) + Cl^-(aq)$$
0 0 +1 -2 +5 -2 +1 -1

I has increased in oxidation number, so oxidation has occurred. Cl has decreased in oxidation state, so reduction has occurred. Since both oxidation and reduction have occurred, this is a plausible reaction and using the rules that follow, we can balance the equation.

7.B. Balancing oxidation-reduction reactions.

It is sometimes helpful to remember that oxidation is **l**oss **o**f **e**lectrons (Leo the lion) and reduction is a gain of electrons (GR). Use the rules and method suggested in the text for balancing equations.

Example 4.8. Balance the equations for the following redox reaction in acidic solution.

$MnO_4^- + C_2O_4^{2-} + H^+ \text{-->} Mn^{2+} + CO_2(g) + H_2O(l)$

Solution:

Step 1. Identify the species undergoing oxidation and reduction. To do this, examine the oxidation state of each atom. Mn goes from +7 to +2, a reduction in O.S. Oxygen stays the same. C goes from an oxidation state of +3 to +4, a gain in oxidation state. We have identified both the oxidation and reduction parts of this reaction. The half reactions are:

$MnO_4^- \text{-->} Mn^{2+}$ (not balanced, oxidation number decrease of 5/Mn)

$C_2O_4^{2-} \text{-->} CO_2$ (not balanced, oxidation number increase of 1/C)

Step 2. Add coefficients so total increase in oxidation number equals total decrease in oxidation number.

$MnO_4^- + 5C_2O_4^{2-} + H^+ \text{-->} Mn^{2+} + 10CO_2(g) + H_2O(l)$

We essentially multiply the increased species by 5 on the left and 10 on the right to keep the C balanced. Because the C is two in $C_2O_4^{2-}$, we can look at the total change as 10 and therefore multiply Mn by 2.

$2MnO_4^- + 5C_2O_4^{2-} + H^+ \text{-->} 2Mn^{2+} + 10CO_2(g) + H_2O(l)$

To balance the O atoms, we add 8 H_2O to the right side of the equation:

$2MnO_4^- + 5C_2O_4^{2-} + H^+ \text{-->} 2Mn^{2+} + 10CO_2(g) + 8H_2O(l)$

Finally, to balance the H, we add $16H^+$ to the left side of the reaction.

$2MnO_4^- + 5C_2O_4^{2-} + 16H^+ \text{-->} 2Mn^{2+} + 10CO_2(g) +8\ H_2O(l)$

Checking, we see that the charge is balanced on each side of the equation. The atoms are also balanced.

8. Substances undergoing both oxidation and reduction.

Disproportionation reactions are reactions in which the different atoms of the same substance undergo oxidation and reduction simultaneously.

9. Oxidizing and reducing agents.

Identify these in reactions and know which metals are the strongest reducing agents. A substance that is oxidized in an oxidation-reduction reaction is a reducing agent. A substance that is reduced is an oxidizing agent. Know the maximum oxidation states of nonmetals and how nonmetals react in oxidation-reduction reactions.

Example 4.9. What would you predict would happen if Cu(s) was added to a solution containing Ag^+ ?

Solution: Cu lies above Ag in the activity series. Therefore, it will displace Ag^+ ions.

$$Cu(s) + 2Ag^+(aq) \rightarrow Cu^{2+}(aq) + 2Ag(s)$$

10. Some practical applications of oxidation and reduction reactions.

In analytical chemistry, $KMnO_4$ is used as an oxidizing agent to determine the percent iron in iron ore. Organic chemists use oxidizing agents to convert ethanol to acetaldehyde and acetic acid. Oxygen is widely used as an oxidizing agent in industrial processes such as welding. There are many other examples of oxidation and reduction reactions.

Exercise 4.12A (text). Suppose the titration described in Example 4.12 were carried out with 0.02250 M $K_2Cr_2O_7$ rather than 0.02250 M $KMnO_4$. What volume of $K_2Cr_2O_7$ would be required?

$$6\,Fe^{2+} + Cr_2O_7^{2-} + 14\,H^+ \text{-->} 6\,Fe^{3+} + 2\,Cr^{3+} + 7\,H_2O$$

Solution: The difference from Example 4.12 is the stoichiometry of 6 mol Fe^{2+}/1 mol $Cr_2O_7^{2-}$. The reaction will therefore take less $K_2Cr_2O_7$ and the amount required is 5/6 that of the $KMnO_4$. The amount required is 5/6 x 26.45 mL = 22.04 mL.

Skills Test Problem

Typically, hard water contains about 50 mg Ca^{2+} per liter. There are many ways to remove Ca^{2+} from hard water, thereby softening it. Sodium carbonate can be added to precipitate the calcium, which can then be filtered off.

$$Ca^{2+}(aq) + CO_3^{2+}\ (aq)\ ?\ CaCO_3\ (s)$$

What mass of sodium carbonate (Na_2CO_3) must be added to soften 1.0 L of hard water (assume the precipitation reaction goes to completion).

Solution

First, calculate the number of moles of calcium present.

$$\frac{50\,\text{mg}}{\text{L}} \times \frac{1\,\text{g}}{1000\,\text{mg}} \times \frac{1\,\text{mole}}{40.08\,\text{g}} \times 1.0\,\text{L} = 0.0012\,\text{moles} = 1.2 \times 10^{-3}\ \text{moles}$$

From the above chemical equation, 1 mole of sodium carbonate will be required for each mole of calcium. Therefore:

$$1.2 \text{ x } 10^{-3} \text{ moles of } Na_2CO_3 \text{ x } \frac{106\,\text{g}}{1\,\text{mole}} = 0.13\ \text{g}$$

Quiz A

1. Which of the following are strong electrolytes?
   a. HCOOH
   b. NH_3
   c. HCl
   d. CH_3OH
2. Which of the following solutions has the highest concentration of Cl^-?
   a. 0.010 M NaCl
   b. 0.006 M $MgCl_2$
   c. 0.005 M $AlCl_3$
   d. 0.012 M KCl
3. What happens to the oxidation state of one of its elements when a compound is oxidized?
4. What is the oxidation state of chlorine in ClO_2^-?
5. Predict whether a reaction is likely to occur, and if so, write the net ionic equation: $CH_3COOH + HNO_3$ -->
6. How does a strong acid differ from a weak acid? Give an example of each.
7. Calculate the volume, in milliliters, of 0.025 M HCl required to titrate 50.00 mL 0.0115 M $Ca(OH)_2$.
8. Define nonelectrolyte. Give an example.
9. How many ionizable hydrogen atoms are present in CH_3CH_2COOH?
10. Write the ionization reaction for ammonia, NH_3.

Quiz B

1. Complete and balance the following half-reaction in base and indicate whether oxidation or reduction is involved. BrO^- --> Br_2
2. What happens to the oxidation state of one of its elements when a compound acts as a reducing agent?
3. What is the oxidation state of carbon in CO_2?
4. Predict whether a reaction is likely to occur, and if so, write a net ionic equation: $BaS + CuSO_4$ -->
5. Which of the following has the highest concentration of NO_3^-?
   a. 0.010 M HNO_3
   b. 0.005 M $Al(NO_3)_3$
   c. 0.010 $LiNO_3$
   d. 0.008 M $Ca(NO_3)_2$

6. Which of the following is a weak electrolyte?
   a. NaCl
   b. $MgCl_2$
   c. CH_3COOH
   d. CH_4
7. Calculate the volume, in milliliters, of 0.015 M NaOH required to titrate 10.00 mL 0.005M HCl.
8. $MgCl_2$ is mixed with NaOH. State whether a reaction occurs, and if so, write the net ionic equation for the reaction that occurs.
9. What is the oxidation state of Ca in $Ca_3(PO_4)_2$?
10. What are the oxidation states of oxygen and fluorine in OF_2?

Quiz C

Multiple Choice

1. A strong electrolyte
   a. is a strong acid or base
   b. undergoes oxidation
   c. produces H^+ in water
   d. nearly completely dissociates in water at appreciable concentration
2. An oxidizing agent
   a. causes oxidation to occur
   b. causes reduction to occur
   c. completely dissociates in water
   d. is usually a solid
3. A weak acid
   a. is a strong electrolyte
   b. undergoes oxidation
   c. is completely dissociated in water
   d. produces H^+ in water
4. In a precipitation reaction
   a. an acid reacts with a base to produce a salt and water
   b. a solid precipitate forms and drops out of solution
   c. transfer of electrons between reactants occurs
   d. products are strong electrolytes

True or False

5. Oxidation is the loss of one or more electrons.
6. HBr is a strong acid.

7. The oxidation state of S in SF_6 is -6.
8. An atom in an uncombined element has an oxidation state equal to the group number.
9. The oxidation state of Mn in $MnCl_2$ is larger than the oxidation state of Na in NaCl.
10. NH_3 is a weak base.

Self Test

1. Identify the following as strong electrolytes, weak electrolytes or nonelectrolytes:
   a. NaBr b. K_2SO_4 c. CH_3COOH d. H_2O e. CH_3OH
2. Predict the solubility of the following compounds:
   a. CdS b. KNO_3 c. $PbCl_2$ d. $BaSO_4$
3. Write the balanced ionic equation and net ionic equation for the following reaction:
   $Ca(OH)_2 + CH_3COOH \rightarrow$
4. Write the chemical formulas for the substances involved in the following reaction and balance the equation:
   Calcium hydroxide + nitric acid → calcium nitrate + water
5. What volume of 0.214 M KOH is required to neutralize 25.00 mL of 0.0500 M HCl?
6. What volume of 6.0 M HCl is required to neutralize 10.0 mL of 0.10 M $Ba(OH)_2$?
7. Predict whether a precipitate will form in the following reactions and if so write the net ionic equation:
   a. $K_2SO_4\ (aq) + NaCl\ (aq) \rightarrow$
   b. $CaCl_2\ (aq) + BaS\ (aq) \rightarrow$
   c. $MgCl_2\ (aq) + Na_2CO_3 \rightarrow$
8. Write oxidation numbers for each element in the following species
   a. NH_3 b. CH_4 c. SO_4^{-2} d. H_2O_2
9. Balance the following equation
   $Fe_2O_3\,(s) + C(s) \rightarrow Fe(s) + CO_2(g)$
10. In the above equation, does oxidation and reduction occur? If so identify the oxidized and reduced species. Write the oxidation number for each species.
11. Predict what if any reaction will occur when Cu^{2+} is mixed with Zn(s). Justify your answer.
12. Which of the following will react with water to release hydrogen gas?
    a. K b. Fe c. Na d. Pt
13. Without doing detailed calculations, which of the following has the highest Na^+ concentrations?
    a. 0.010 M NaCl
    b. 0.050 g of NaCl in 1.0 L of solution

c. 0.05 moles of NaCl in 2.0 L of solution

d. 0.010 moles of NaCl in 100 mL of solution

14. Are all hydrogen containing compounds acids? Explain.
15. What happens to a reducing agent in an oxidation-reduction reaction?
16. Can oxidation occur without reduction being present? Vitamin E is said to be an anti-oxidant. What would you predict about its ability to be oxidized or reduced?

CHAPTER 5

Gases

This chapter will describe some of the basic properties of gases and the related theory to explain their behavior. We will use the ideal gas law to calculate the relationships between pressure, temperature, volume and amount of gases. Kinetic Molecular Theory will be used to explain many of the properties of gases.

Chapter Objectives: This chapter should enable you to:

1. Describe some of the general properties of gases.
2. Use the Kinetic Molecular Theory to explain physical properties of gases on the molecular level.
3. Define pressure and convert various pressure units.
4. Calculate the pressure exerted by a column of liquid and describe how this can be used to determine atmospheric pressure in a manometer.
5. Use Boyle's law to relate pressure and volume of a gas.
6. Use Charles' law to relate temperature and volume of a gas.
7. State Avogadro's law and the define STP.
8. State the Ideal Gas Law and know how to rearrange and use the equation.
9. Use the mass of a gas and the Ideal Gas Law equation to determine the molecular weight of a gas.
10. Calculate the density of gases.
11. State Gay-Lussac's Law of Combining Volumes and use this to calculate volumes of gases in reactions.
12. Determine the gas stoichiometry in reactions using gas volumes.
13. Apply Dalton's law of partial pressures to determine the total pressure of a mixture of gases or partial pressures of individual gases.
14. State the basic postulates of the Kinetic-Molecular Theory.
15. Use Kinetic-Molecular Theory to relate temperature and molecular motion.
16. State Graham's Law and use this principle to calculate the relative rates of effusion for two different gases.
17. Describe the conditions under which gases are most likely to be nonideal.

Chapter Summary

1. General properties of gases

How do gases differ from liquids and solids in terms of density?

Gases are composed of molecules in constant motion. Gases are mostly empty space because the molecules are far apart. The density of gases is much smaller than the density of liquids or solids; for many gases at 25 °C and 1 atm the density is about 1/1000 that of water.

2. Kinetic Molecular Theory was developed to provide a model for gases on the molecular level and to explain their observed physical properties.

3. Defining pressure and the units used to describe pressure in the SI system. Converting one set of units to another.

Pressure = Force/Area

The SI unit of pressure is the Pascal, abbreviated Pa. Other units for pressure are the atmosphere (1 atm = 101,325 Pa) and mmHg or torr.

1 atm = 760 mm Hg = 760 torr

Example 5.1. Carry out the following conversions.

a. 722 torr to mm Hg

Solution: The conversion here is 760 mm Hg = 760 torr or 1mm Hg = 1 torr. Therefore, 722 torr x 1 mmHg/1 torr = 722 mmHg

b. 98.2 kPa to torr

Solution: From the above relationships, we know that 101.325 kPa = 1 atm = 760 torr. Therefore 760 torr = 101.325 kPa or 1 torr = 0.133322 kPa.

$$98.2 \text{ kPa x } \frac{1 \text{ torr}}{0.13332 \text{ kPa}} = 737 \text{ torr or } 98.2 \text{ kPa x } \frac{760 \text{ torr}}{101.325 \text{ kPa}} = 737 \text{ torr}$$

If we check, we notice that 98.2 kPa is slightly less than 1 atm and 737 torr is also slightly less than 1 atm. Our estimation agrees with the answer.

c. 29.95 in Hg to torr

Solution: Here we can do multiple conversions to arrive at the answer. If we convert from in. Hg to mm Hg the conversion to torr is trivial.

$$29.95 \text{ in Hg x } \frac{2.54 \text{ cm}}{1 \text{ in.}} \text{ x } \frac{10 \text{ mm}}{1 \text{ cm}} = 760.7 \text{ mm Hg} = 760.7 \text{ torr}$$

d. 768 torr to atm

Solution: 1 atm = 760 mmHg = 760 torr. Therefore 1 atm/760 torr is the conversion factor to use

$$768 \text{ torr x } \frac{1 \text{ atm}}{760 \text{ torr}} = 1.01 \text{ atm}$$

4. Calculate the pressure exerted by a column of liquid and describe how this can be used to determine atmospheric pressure in a manometer.

The pressure exerted by a column of liquid is

$P = g \times d \times h$

where P is the pressure, g is the acceleration due to gravity (9.807 m s^{-2}), d is the density of the liquid, and h is the height of the column.

A manometer measures gas pressure in a closed container. The difference between barometric pressure and the pressure exerted by the gas is determined from the difference in mercury levels in the open and closed ends of the manometer.

Example 5.2. Calculate the height of a column of carbon tetrachloride, CCl_4, that would exert the same pressure as a column of mercury that is 760 mm high. The density of CCl_4 is 1.59 g/cm^3 and that of Hg is 13.6 g/cm^3.

Solution: The formula for the pressure exerted by a column of liquid is

$$P = g \times d \times h$$

where g is the force of gravity, d is the density of the liquid and h is the height of the column. We can solve this problem by first calculating the height of the Hg column (if we know g) and then calculating the height of the carbon tetrachloride column, but this is cumbersome. Since we know the pressure exerted by two columns will be equal, we can relate the two equations and solve for the height.

$$P_{Hg} = g \cdot d_{Hg} \cdot h_{Hg} = P_{CCl_4} = g \cdot d_{CCl_4} \cdot h_{CCl_4}$$

rewriting we have

$$g \cdot d_{Hg} \cdot h_{Hg} = g \cdot d_{CCl_4} \cdot h_{CCl_4}$$

Cancelling g from both sides and dividing by d_{CCl4}, we have the height of the CCl_4 column:

$$h_{CCl_4} = \frac{d_{Hg} \cdot h_{Hg}}{d_{CCl_4}} = \frac{\frac{13.6\text{ g}}{\text{cm}^3} \times 760\text{ mm Hg}}{\frac{1.59\text{ g}}{\text{cm}^3}} = 6500\text{ mm}$$

Notice the relationship between the height of the column and the density of the liquid. Since CCl_4 is about 1/10 as dense as Hg we would expect the height of the column to be about 10 times higher. This is approximately what we get. Why don't we use water barometers?

5. Boyle's Law: Relate pressure and volume for a gas.

At constant temperature the product of pressure x volume is a constant. This means that if we have the same amount of gas at the same temperature but change the volume, we can calculate the new pressure. This is simply stated as:

$$P_1V_1 = P_2V_2$$

Exercise (text) 5.4A. A sample of helium occupies 535 mL at 988 torr and 25° C. If the sample is transferred to a 1.05-L flask at 25° C, what will be the gas pressure in the flask?

First, notice that the temperature is the same. Therefore, we can use the above relationship for pressure and volume. Next notice that the initial and final volumes are not in the same units. Converting mL to L (or vice versa) we can calculate the new pressure.

$$988 \text{ torr} \times 0.535 \text{ L} = P_2 \times 1.05 \text{ L}$$

Solving for P_2 by dividing both sides by 1.05 L gives:

$$P_2 = \frac{988 \text{ torr} \times 0.535 \text{ L}}{1.05 \text{ L}} = 503 \text{ torr}$$

Notice that since the volume has almost doubled, P should be reduced by almost half. This is an inverse relationship: As one quantity increases the other decreases.

6. Charles' Law: Relate temperature and volume for a gas.

A similar relationship to Boyle's Law can be written for temperature and volume. In this case, however, the two quantities are directly proportional. That is, for a fixed amount of gas, as the temperature increases the volume also increases. (Try heating a balloon.) This can be written as:

$$\frac{V_1}{T_1} = \frac{V_2}{T_2}$$

For this relationship to be valid, we MUST use the Kelvin temperature scale.

Example 5.3. A balloon at room temperature (25° C) has a volume of 1.5 L. The balloon is now heated to 100 °C. What is the volume of the hot balloon?

First convert to the Kelvin temperature scale.

$$T_1 = 25\ °\text{C} + 273 = 298 \text{ K}$$
$$T_2 = 100\ °\text{C} + 273 = 373 \text{ K}$$

Now relating the two temperatures and one volume, we can solve for the other volume:

$$\frac{1.5\text{L}}{298\text{K}} = \frac{V_2}{373\text{K}}$$

Multiplying both sides by 373 K, we get V_2.

$$V_2 = \frac{1.5 \text{ L}}{298 \text{ K}} \text{x} 373 \text{ K} = 1.9 \text{ L}$$

Notice what happens if you forget to convert the temperature to K. The volume will be much too large.

Example 5.4. A sample of gas occupies 645 L at 615 °C. What is the temperature of the gas when it occupies 200 L, assuming the pressure is constant?

Solution: In this case we have two volumes and one temperature so we must solve for the other temperature. The same relationship above works, but we must do the algebra.

$$\frac{V_1}{T_1} = \frac{V_2}{T_2}$$

Rearranging, we multiply by T_2 to give:

$$\frac{T_2 \times V_1}{T_1} = V_2$$

Now multiply by T_1 and divide by V_1.

$$T_2 = \frac{V_2 \times T_1}{V_1}$$

We're almost there, but we need to remember to convert to Kelvin.

$T_1 = 615\ °C + 273\ K = 888\ K.$

$$T_2 = \frac{200\ L \times 888\ K}{645\ L} = 275\ K.$$

We have reduced the volume by about a third, so the temperature, should also be about a third. When dealing with equations such as the one above and solving for an unknown quantity, it is usually easiest to move the quantity you want to the numerator. Next multiply and divide by the other terms to isolate the term you are solving for.

7. Avogadro's Law. Standard conditions of temperature and pressure (STP) and the molar volume of a gas.

An important principle, Avogadro's Law, is that a given amount of **any** gas will occupy the same volume at the same temperature and pressure. The standard conditions of temperature and pressure are 273.15 K and 1 atm (760 torr). One mole of gas will occupy 22.4 L under these conditions. We can use volumes to calculate stoichiometry in the same way we use moles if the products and reactants are gases and T and P stay the same. We can also use this relationship to calculate the mass of gas at STP or the volume of a given mass of gas at STP.

Example 5.5. What is the mass of 2.00 L of $CO_2(g)$ at STP?

Solution: Using the relationship above, we know that one mole occupies 22.4 L. Using this as a conversion factor, we can calculate the mass.

$$2.00\ L \times \frac{1\ \text{mole}\ CO_2}{22.4\ L} = 0.0893\ \text{mol}$$

$$0.0893\ \text{mole} \times \frac{44.0\ g}{1\ \text{mol}\ CO_2} = 3.93\ g\ CO_2$$

Example 5.6. What is the volume of 6.00 g of CH_4 at STP?

Solution: Here we convert from mass to volume. First we convert to moles and then to volume.

$$6.00\text{ g x } \frac{1\text{ mole }CH_4}{16.04\text{ g}} \text{ x } \frac{22.4\text{ L}}{1\text{ mole}} = 8.38\text{ L}$$

The combined gas law is

$$\frac{P_1V_1}{n_1T_1} = \frac{P_2V_2}{n_2T_2}$$

and is most useful when a gas is described under an initial set of conditions and a question is asked about the final conditions. If you know the combined gas law, you can easily obtain Boyle's, Charles', and Avogadro's Laws from it.

Example 5.7. Two containers of identical volume at the same temperature are filled with gases. Container A contains twice as many moles of gas as does container B. What is the ratio of the pressures of the two gases?

Solution: Since $T_1 = T_2$ and $V_1 = V_2$, then $\frac{P_1}{n_1} = \frac{P_2}{n_2}$ and since $n_1 = 2n_2$ then $P_1 = 2P_2$. The pressure in container A is twice the pressure in container B.

8. The Ideal Gas Law

The ideal gas law combines the previous relationships among P,V, T and n to give the relationship

$$PV = nRT$$

where R is the gas constant and n is the number of moles of gas.

Example 5.8. What is the amount of nitrogen, in mol N_2, in a sample that occupies 35.0 L at a pressure of 3.15 atm and a temperature of 852 K?

Solution: We need to rearrange the Ideal Gas Law to solve for the amount, n, of gas:

$$n = \frac{PV}{RT} = \frac{3.15\text{ atm x }35.0\text{L}}{0.0821\text{ L atm mol}^{-1}\text{ K}^{-1}\text{ x }852\text{ K}}$$

$$= 1.58\text{ mol}$$

Example 5.9. What is the temperature at which 20.0 g of H_2 gas will exert a pressure of 680 torr in a volume of 350.0 L?

Solution: Rearrange the Ideal Gas Law to solve for temperature.

$$T = \frac{PV}{nR}$$

Next convert the quantities given to appropriate units. First convert 20.0 g to moles:

$$20.0 \text{ g } H_2 \times \frac{1 \text{ mol } H_2}{2.02 \text{ g}} = 9.90 \text{ mol } H_2$$

$$T = \frac{\frac{680}{760} \text{ atm} \times 350.0\text{L}}{9.90 \text{ mol} \times 0.0821 \text{ L atm mol}^{-1}\text{K}^{-1}} = 385 \text{ K}$$

Note that P was converted to atm and that the resulting calculation is in Kelvin.

9. Use the mass of a gas and the Ideal Gas Law equation to determine the molecular weight of a gas.

By using the definition of molar mass, *M*

$$M = \frac{\text{mass of substance}}{\text{moles of substance}} = \frac{m}{n}$$ and the ideal gas equation

$$n = \frac{PV}{RT} = \frac{m}{M}$$ we can arrive at an equation using the molar mass

$$M = \frac{mRT}{PV}$$

Example 5.10. Calculate the molar mass of a gas, 0.440 g of which occupies 180.0 mL at 562.4 mm Hg and 86 °C.

Solution: Calculate the amount of gas from the Ideal Gas Law. Remember to convert to the appropriate units.

$$n = \frac{PV}{RT} = \frac{\frac{562.4}{760} \text{ atm} \times 0.1800 \text{ L}}{0.0821 \text{ L atm mol}^{-1} \text{ K}^{-1} \times 359.15 \text{ K}} = 0.00452 \text{ mol}$$

To find the molar mass of the gas, use the definition:

$$M = \frac{m}{n} = \frac{0.440 \text{ g}}{0.00452 \text{ mol}} = 97.3 \text{ g/mol}$$

10. Density of gases

The density is defined as mass/volume. We can use this definition and the Ideal Gas Law to calculate the density of gases under different conditions. Rearranging from the ideal gas equation using molar mass,

$$M = \frac{mRT}{PV} \qquad \frac{m}{V} = M \times \frac{P}{RT} = d$$ where d is the density

Exercise (text) 5.14A. Calculate the density of ethane gas (C_2H_6), in g/L, at 15 °C and 748 torr.

Solution: Use the relationship between density and the Ideal Gas Law:

$$d = M \times \frac{P}{RT}$$

We use this relationship because we can calculate the molar mass of ethane from the periodic table. Adding the appropriate atomic weights, we get the molar mass of ethane = 30.07 g/mol. Now we can convert the P and T to atm and K, respectively. P = 748/760 atm and T = 15 °C + 273.15 = 288 K. Substituting into the above equation:

$$d = \frac{30.07\,\text{g mol}^{-1} \times 0.984\,\text{atm}}{0.0821\,\text{L atm mol}^{-1}\,\text{K}^{-1} \times 288\,\text{K}} = 1.25\text{ g/L}$$

Exercise (text) 5.15A. To what temperature must propane gas (C_3H_8) at 785 torr be heated to have a density of 1.51 g/L?

Solution: In this case we know the density and must solve for temperature. Rearranging the above equation gives:

$$T = M \times \frac{P}{Rd}$$

The strategy is now to calculate the molar mass from the atomic weights of the elements. M = 44.10 g/mol.

Substituting into the above equation:

$$T = \frac{44.10\,\text{g mol}^{-1} \times \frac{785}{760}\,\text{atm}}{0.0821\,\text{L atm mol}^{-1}\,\text{K}^{-1} \times 1.51\,\text{g L}^{-1}} = 367\text{ K}$$

11. Gay-Lussac's Law of Combining Volumes and the volumes of gases in reactions

A simple statement of Gay-Lussac's Law is that gases at the same temperature and pressure will combine in small whole number ratios. These small whole number ratios will also give the coefficients of the chemical equations involved.

Example 5.11. What are the combining volume ratios for the reaction:

$$2\,CO(g) + O_2(g) \text{ --> } 2\,CO_2(g)$$

Solution: The ratios are 2:1:2. Note that we must recognize that oxygen is a diatomic molecule to get the correct ratios.

Example 5.12. In the above reaction, what volume of oxygen must react to produce 5.0 L of CO_2?

Using the above ratios, we know that one volume of oxygen is required for each two volumes of CO_2.

$$5.0 \text{ L of } CO_2 \times \frac{1 \text{ L } O_2}{2 \text{ L } CO_2} = 2.5 \text{ L of } O_2 \text{ required.}$$

Example 5.13. What volume of oxygen is required to burn 0.556 L of propane, C_3H_8 ?

$$C_3H_8(g) + 5\ O_2(g) \text{ --> } 3\ CO_2(g) + 4\ H_2O(g)$$

Assume all gases are at the same temperature and pressure.

Solution: First, check that the equation is balanced. We should have the same number of atoms of each element on each side of the equation. In this case the equation is already balanced.

Second, from the balanced equation we can obtain the volume ratios of reactants and products. In this case the ratio of C_3H_8:O_2 is 1:5. For each molecule of propane, five molecules of O_2 are required for complete combustion. This means that for each volume of propane, five volumes of oxygen are required. Therefore, to calculate the volume of oxygen:

$$0.556 \text{ L} \times \frac{5 \text{ L } O_2}{1 \text{ L } C_3H_8} = 2.78 \text{ L}$$

Similarly, we can calculate the volume of the products using the same equation. In this case the ratio of C_3H_8:CO_2 is 1:3. Multiplying 0.556 L by 3 will give the volume of CO_2 produced. We can also calculate this from the volume of O_2. In this case the ratio is 5 O_2: 3 CO_2. To calculate the amount of CO_2, multiply the amount of oxygen by 3/5. These problems can be solved in the same way as conversion factor problems. The ratio is written so that the known species cancels and the remaining species is the quantity that is desired to be calculated.

12. Stoichiometry in reactions using gas volumes.

Gay-Lussac's Law applies only if the gases are at the same temperature and pressure. Since identical amounts of two gases will occupy the same volumes under identical T and P conditions, we can relate volumes in a reaction. This actually relates the amount in moles. Using the ideal gas equation:

$$n_1 = \frac{P_1V_1}{RT_1} \quad \text{and} \quad n_2 = \frac{P_2V_2}{RT_2}$$

If $P_1 = P_2$ and $T_1 = T_2$, then

$$n_2 = \frac{P_1V_2}{RT_1} \quad \text{and} \quad n_1 = \frac{P_1V_1}{RT_1}$$

The number of moles is directly proportional to the volume since each volume is multiplied by the same constant.

Exercise (text) 5.17A. The manufacture of quicklime (CaO) for use in the construction industry is accomplished by decomposing limestone ($CaCO_3$) by heating.

$CaCO_3(s)$ --> $CaO(s) + CO_2(g)$

What volume of $CO_2(g)$ at 825 °C and 754 torr is produced in the decomposition of 45.8 kg $CaCO_3(s)$?

Solution: First the equation above gives the moles of $CO_2(g)$ formed for each mole of $CaCO_3$. The ratio is one to one. We therefore need to calculate the number of moles of $CaCO_3$ used. This will give the number of moles of CO_2 from the reaction stoichiometry. We can then use the Ideal Gas Law to determine the volume of gaş produced.

Step 1. $45.8\,\text{kg CaCO}_3 \times \frac{1000\,\text{g}}{\text{kg}} \times \frac{1\,\text{mole}}{100.1\,\text{g}} = 458\,\text{mol CaCO}_3$

$$458\,\text{mol CaCO}_3 \times \frac{1\,\text{mol CO}_2}{1\,\text{mol CaCO}_3} = 458\,\text{mol CO}_2$$

Step 2. Calculate the volume from the Ideal Gas Law:

$$V = \frac{nRT}{P}$$

$$V = \frac{458\,\text{mole} \times 0.0821\,\text{L atm mol}^{-1}\,\text{K}^{-1} \times 1098\,\text{K}}{0.992\,\text{atm}} = 4.16 \times 10^4\,\text{L}$$

Example 5.14. What mass of CaO(s) is formed in the decomposition of $CaCO_3(s)$ if the volume of $CO_2(g)$ obtained is 1.25×10^4 L at 825 °C and 733 torr?

$CaCO_3(s)$ --> $CaO(s) + CO_2(g)$

Solution: We know V, P and T for the gas, so we can calculate the number of moles. Using the above reaction stoichiometry, we know that for each mole of CO_2 formed, one mole of CaO was also formed.

$$n = \frac{PV}{RT} = \frac{0.964\,\text{atm} \times 1.25 \times 10^4\,\text{L}}{0.0821\,\text{L atm mol}^{-1}\,\text{K}^{-1} \times 1098\,\text{K}} = 134\,\text{mol}$$

$$\text{Mass of CaO} = 134\,\text{mol} \times 56.08\,\text{g/mol} = 7.51 \times 10^3\,\text{g}$$

13. Dalton's Law of Partial Pressures: Applying this law to determine the total pressure of a mixture of gases or partial pressures of individual gases.

Dalton's Law of Partial Pressures states that the total pressure of a mixture of gases is equal to the sum of the partial pressures exerted by the separate gases. We can understand this idea from the postulates of Kinetic Molecular Theory. Since the particles in an ideal gas do not interact, it is only the total number of particles that is important, not the identity of the individual particles.

Exercise (text) 5.18B. What is the total pressure exerted by a mixture of 4.05 g N_2, 3.15 g H_2, and 6.05 g He when confined to a 6.10 L container at 25 °C?

Solution: Since the total pressure is the sum of the partial pressures and since the pressure is related to the number of moles of gas, we can work this problem in at least two ways. Probably the easiest and most straightforward is to calculate the number of moles of each gas. The total pressure is then $n_{total} = PV/RT$, where n_{total} is the sum of moles of each gas. We can also calculate the partial pressure for each gas and then total the partial pressures, but this method is more cumbersome. Using the first approach:

$$4.05 \text{ g } N_2 \text{ x } \frac{1 \text{ mol } N_2}{28.01 \text{ g}} = 0.145 \text{ mol}$$

$$3.15 \text{ g } H_2 \text{ x } \frac{1 \text{ mol } H_2}{2.016 \text{ g}} = 1.56 \text{ mol}$$

$$6.05 \text{ g He x } \frac{1 \text{ mol He}}{4.003 \text{ g}} = 1.51 \text{ mol}$$

$$n_{tot} = 0.145 \text{ mol} + 1.56 \text{ mol} + 1.51 \text{ mol} = 3.22 \text{ mol}$$

Using the Ideal Gas Law for pressure,

$$P = \frac{nRT}{V} = \frac{3.22 \text{ mole x } 0.0821 \text{ L atm mol}^{-1}\text{K}^{-1} \text{x } 298.15\text{K}}{6.10 \text{ L}} = 12.9 \text{ atm}$$

Example 5.15. A sample of expired air is composed, by volume, of the following main components: N_2, 74.1%; O_2, 15.0%; H_2O, 6.0%; Ar, 0.9%; and CO_2, 4.0%. What are the partial pressures of each of the five gases in the expired air at 37 °C and 1.000 atm?

Solution: We need to relate partial pressure to partial volume. The relationship to use is:

$$\frac{P_1}{P_{tot}} = \frac{V_1}{V_{tot}}$$

Rearranging,

$$P_1 = \frac{V_1 P_{tot}}{V_{tot}}$$

This can be applied for each component.

We are not given the total volume; however, the volume percent is the volume occupied by each component in 100 L of air; for N_2 the partial volume is 74.1 L, and so on for the other components. Therefore,

$$\frac{V_1}{V_{tot}} \text{x} 100 = \% \text{ volume}$$

For N_2, $$\frac{P_{N_2}}{P_{tot}} = \frac{V_{N_2}}{V_{tot}} = \frac{\%\,\text{volume}}{100} = 0.741$$

$P_{N2} = 0.741 \times P_{tot} = 0.741 \times 1.00\ \text{atm} = 0.741\ \text{atm}$

For each of the other components, the same calculation applies.

14. The basic postulates of the Kinetic-Molecular Theory

The postulates of kinetic-molecular theory as they relate to particles of a gas can be stated as follows:

1. The volume of the particles is small compared to the total volume of the gas.
2. The particles of the gas do not interact with each other. There are no strong attractive or repulsive forces between the gas molecules.
3. The particles are in constant motion and collisions are elastic. This explains why gases exert pressure on a container.
4. The average translational kinetic energy of the particles is directly related to the temperature.

15. Use Kinetic-Molecular Theory to relate temperature and molecular motion.

The average translational kinetic energy of a gas is directly proportional to the Kelvin temperature.

$$\overline{KE} = \frac{3}{2} \cdot \frac{R}{N_A} \cdot T$$

Example 5.16. Which of the following gases under the same conditions of pressure and temperature has the largest kinetic energy?

a. H_2 **b.** Cl_2 **c.** Ar **d.** All are the same

Solution: Since the translational kinetic energy is proportional to a constant x T, all gases have the same kinetic energy and it does not depend on the mass of the gas.

Example 5.17. Which of the following gases under the same conditions of pressure and temperature has the highest root-mean-square velocity?

a. H_2 **b.** Cl_2 **c.** Ar **d.** All are the same

Solution: The root-mean-square speed is directly proportional to temperature and inversely proportional to molecular mass. Therefore, the smallest molar mass will have the highest speed. Answer *a*, H_2 is correct.

16. Effusion and Diffusion

Effusion is the passage of a gas through a tiny opening into an evacuated chamber. Diffusion is the mixing of gases. Because gas molecules are in constant motion, we would expect that when two gases are mixed, the particles will become randomly distributed in the container.

Graham's Law and calculating the relative rates of effusion for two different gases.

Graham's Law of effusion states that at a given temperature, the rates of effusion of gas molecules are inversely proportional to the square roots of their molar masses.

$$\frac{\text{Rate}_1}{\text{Rate}_2} = \sqrt{\frac{M_2}{M_1}}$$

Since the lighter gases have a higher average velocity, we would expect them to effuse more quickly.

Example 5.18. Calculate the relative rates of effusion of N_2 and Ar under the same conditions of temperature and pressure.

Solution: Since we know that lighter gases move faster, we expect that the rate of effusion of nitrogen would be greater than argon. Graham's Law gives us:

$\frac{\text{rate 1}}{\text{rate 2}} = \sqrt{\frac{M_2}{M_1}} = \sqrt{\frac{N_2}{Ar}} = \sqrt{\frac{28.01}{39.948}} = 0.84374$ or N_2 has an effusion rate about 1.2 times greater than Ar.

17. Conditions of nonideal gas behavior

Gases are more likely to be nonideal under conditions of low temperature and higher pressure. Under these conditions, the volume of the particles of a gas is less likely to be negligible compared to the total volume and the particles are more likely to interact. An extreme example of nonideal gas behavior is the liquification of gases at low temperature and high pressure, such as liquid nitrogen.

Van der Waals equation for real gases.

The van der Waals equation is

$$\left(P + \frac{n^2 a}{V^2}\right)(V - nb) = nRT$$

The term $\frac{n^2 a}{V^2}$ is added to the pressure to account for the intermolecular forces of attraction. The particles attract each other and reduce the pressure from what would be expected for an ideal gas.

The term -nb accounts for the volume of the particles themselves. It is subtracted to give the volume actually available to a given gas. The van der Waals constants, a and b, will be different for each gas.

Skills Test Problem

During World War II, it became necessary to separate $^{235}_{92}U$ from the more abundant isotope $^{238}_{92}U$. Since the chemical properties are almost identical, a chemical separation was not possible. Effusion of uranium hexafluoride (UF_6) was used. Calculate the relative rates of effusion of the two isotopes of uranium as UF_6.

Solution.

The relative rates of effusion are

$$\frac{\text{Rate 1}}{\text{Rate 2}} = \sqrt{\frac{M_2}{M_1}}$$

$$M_1 = {}^{235}_{92}UF_6 = (235) + 6(19.00) = 349.0 \text{ g/mole}$$
$$M_2 = {}^{238}_{92}UF_6 = (238) + 6(19.00) = 352.0 \text{ g/mole}$$

$$\frac{\text{Rate 1}}{\text{Rate 2}} = \sqrt{\frac{352.0}{349.0}} = 1.004$$

Quiz A

True or False

1. According to kinetic-molecular theory, gases move more slowly at higher temperatures.
2. If the temperature of a gas is increased at constant volume, the pressure will also increase.
3. The molar volume of N_2 will be greater than that of He when both gases are at STP.

Problems/Short Answer

4. Convert 1.90 atm to mmHg and torr.
5. Calculate the mass in grams of O_2 in 2.5 L at 0.75 atm and 60.0° C.
6. What volume of $O_2(g)$ measured at 20°C and 1 atm is consumed in the combustion of 4.0 L of $CH_4(g)$ measured at STP?
7. The SI unit for pressure is ____________________.
8. Why can a gas be much more easily compressed than a liquid?
9. Is the actual volume of a gas at very high pressure different from that predicted by the ideal gas law? If so, how?
10. Does the pressure of a gas in a closed container increase or decrease with temperature?

Quiz B

Assume all gases are under the same conditions of temperature and pressure.

1. Which gas has the highest kinetic energy?

   a. H_2 b. O_2 c. Ar d. All are the same

2. Which gas has the highest root-mean-square speed?

   a. H_2 b. O_2 c. Ar d. All are the same

3. Which gas has the highest density?

   a. H_2 b. O_2 c. Ar d. All are the same

4. Calculate the molecular weight of a gas if 0.871 g occupies 250 mL at 760 mm Hg and 20° C.

5. What mass of limestone, $CaCO_3$, must decompose to produce 4.00 L of CO_2(g) at STP?

   $CaCO_3(s) \rightarrow CaO(s) + CO_2(g)$

6. Calculate the relative rates of effusion for O_2 and Xe under the same conditions of temperature and pressure.

7. Calculate the mass in grams of 500 mL of H_2 at STP.

8. Standard temperature and pressure are defined as _____________.

9. What does Charles' law describe?

Quiz C

Matching

Pressure	a. states the relationship between volume and temperature at constant pressure
Boyle's Law	b. states the relationship between volume and pressure at constant temperatue and amount of gas
Ideal Gas Law	c. describes how the volume of a gas is affected by changes in pressure, temperature and amount of gas
Partial pressures	d. mixing of gases by random molecular motion
Kinetic molecular theory	e. escape of a gas through a tiny hole
Diffusion	f. force exerted per unit area
Effusion	g. pressure exerted by each individual gas which is equal to the pressure that would be exerted if the gas were alone in the container.
Charles' Law	h. a model to explain the behavior of gases.

2. Calculate the pressure, in atmospheres, exerted by a column of $H_2O(l)$ that is 2.8 m high. (Hg d = 13.6 g/mL)

3. A sample of gas at 20°C and 1 atm of pressure is heated at constant pressure until its volume doubles. What is the new temperature?

4. Which will diffuse faster, CH_4 or Cl_2 ?
5. Calculate the volume occupied by 50.0 g of O_2 at 25 ^{0}C and 1.0 atm pressure.

Self Test

True-False

1. At fixed pressure, the volume of a gas is inversely proportional to absolute temperature.
2. At fixed temperature, the pressure of a gas is directly proportional to volume.
3. The most compressible state of matter is the gaseous state.
4. A heavier gas (in terms of molecular weight) will diffuse faster than a lighter gas.
5. The SI unit of pressure is the atm.
6. Standard temperature is 298 K.
7. Deviations from the ideal gas law are more pronounced at high pressure.

Problems

8. A mixture of gases contains, by volume: 75.00% N_2,15.00% O_2, and 10.00% Ar. What are the partial pressures of each of the gases in the sample at 1.000 atm?
9. What is the value of u_{rms} of Cl_2 molecules at 0 °C, if u_{rms} of H_2 at 0 °C is 1838 m/s?
10. State the conditions under which a gas is likely to deviate from the ideal gas law.
11. What is the volume of 0.780 mol of Ar at STP?
12. The density of a sample of methane gas was found to be 0.590 g/L at 1 atm pressure. What was the temperature of the sample?
13. What effect will the following changes have on the volume of a gas?
    a. decrease in pressure at constant temperature
    b. increase in the number of moles at constant pressure and temperature
    c. increase in temperature at constant pressure
14. Explain Boyle's law using Kinetic Molecular Theory.
15. The oxidation of ammonia is an important reaction in the production of fertilizer:
    $4\ NH_3\ (g) + 5\ O_2\ (g) \rightarrow 4\ NO\ (g) + 6\ H_2O\ (g)$
    How many liters of NO at 873 K and 1 atm can be produced from 50 L of NH_3 at 373 K and 1 atm?
16. Calculate the ratio of the effusion rates of Ar and Ne from the same container at the same temperature and pressure.

CHAPTER 6

Thermochemistry

Thermochemistry is the study of heat changes that occur during reactions. In order to understand heat changes, we first define different types of energy changes that take place. The First Law of Thermodynamics is introduced and the concept of state functions. We will then use these concepts to calculate energy changes for reactions.

Learning Objectives:

1. Define kinetic and potential energy.
2. Define the following: systems and surroundings, internal energy, heat, and work.
3. State the first Law of Thermodynamics and know how to apply it.
4. Describe what is meant by a state function.
5. Define enthalpy and calculate enthalpy changes.
6. Know how quantities of heat are measured and define heat capacity.
7. Define molar heat capacity and specific heat.
8. State Hess's Law and know how to apply it.
9. Use ΔH_f° to determine the enthalpy changes for chemical reactions.
10. Know how to calculate the enthalpy of ionic reactions.

Chapter Summary

1. Kinetic and potential energy

First, energy is the capacity to do work or produce heat. Kinetic energy may be simply defined as the energy of motion and potential energy may be defined as stored energy (the potential to do work). The SI unit of energy is the Joule and $1\ J = 1(kg\ m^2)/s^2$.

Example 6.1. Which of the following are primarily examples of kinetic energy and which are examples of potential energy?

a. Water behind a dam

b. A car traveling along the highway

c. Fuel in a tank

d. A rocket lifting off from a launch pad

Can something have both potential and kinetic energy?

Solution. (a) and **(c)** have the potential to do work. Water behind a dam can be released to do work and fuel in a tank can be burned to produce heat or to drive a motor to do work. **(b)** and **(d)** are primarily examples of kinetic energy since most of the energy is going toward motion of the object.

Yes, something can have both potential and kinetic energy. A ball that rolls halfway down an incline has kinetic energy and it has potential energy because it has not reached the bottom of the incline.

2. Systems and surroundings, internal energy, heat, and work

When we speak of the **system** in a thermodynamic experiment, we are referring to the portion of the universe we have selected for study. Everything else is the **surroundings**. The **internal energy** of the system is the total energy within the system. You should state these definitions in your own words. Heat is a concept that we have an intuitive understanding of, but it can be difficult to give a definition. The definition of **heat** is the energy transferred from one object to another as a result of a temperature difference. We also have an intuitive understanding of work. We define **work** as a form of energy transfer between a system and its surroundings that can be expressed as force acting through a distance. From these two definitions, it should not be surprising that the energy change for a system is the amount of heat added or given off by the system plus the work done on the system or by the system. We will see later that whether heat or work increase or decrease the energy of a system depends on whether heat (or work) is added to the system (an increase in energy) or given off (done by the system in the case of work).

Example 6.2. Is energy transferred to the system or to the surroundings in the following examples?

a. A block of ice melts.

b. Ammonium nitrate dissolves in water and the solution gets very cold.

c. Ethane burns in oxygen.

d. A gas is compressed.

Solution.

a. Heat is transferred to the ice to convert it from a solid to a liquid.

b. The solution absorbs heat from the surroundings. Immediately after the ammonium nitrate is dissolved, the reaction uses heat and heat must be added to the solution from the surroundings.

c. Heat is transferred to the surroundings. Combustion gives off heat.

d. Compression of a gas involves doing work on the gas. If work is done ON the gas, then the system increases in energy.

3. The First Law of Thermodynamics

There are several ways of stating the First Law of Thermodynamics. One is that the energy of the universe is constant. (State the First Law in your own words.) What does this mean if we are discussing a system that doesn't interact with its surroundings? If the system does not exchange heat or work with its surroundings (an isolated system), and the energy of the universe is constant, then the energy of the system must also be constant. In a system that interacts with its surroundings, energy transfer is measured as heat and work. The energy change for the system can be written as: $\Delta E = q + w$.

If we think of ΔE as $E_{final} - E_{initial}$, which represents the difference in energy between the final state and the initial state of the system, we can understand the signs of q, the heat, and w, the work, transferred. If ΔE is positive, the energy of the system has increased. We would expect that adding heat to a system would increase the energy. Therefore, heat added to the system (absorbed by the system) will be positive. Similarly, heat given off or leaving the system will be negative. It requires energy to do work, so if the system does work the energy of the system decreases. Work done by the system will decrease the energy of the system and will have a negative sign. If work is done on the system, the energy will increase. For example, it requires energy to compress a gas and, therefore, if a gas is compressed, work is done on the system and the energy of the system will increase. Similarly, if a system does work, such as a gas expanding to drive a piston, the system loses energy.

Example 6.3. It takes 89 J of work to compress a certain gas, which gives off 567 J of heat during the process. What is ΔE of the gas?

Solution: Use the equation $\Delta E = q + w$.
Now we must decide on the appropriate signs of q and w. Work is done to compress the gas. We have supplied the work from the surroundings, so the energy of the system will increase by the amount of the work supplied. The work is therefore positive.

$$\Delta E = q + 89\ J$$

The second part is to determine the sign of q. Heat is given off by the gas and the system has lost energy. Therefore, q is negative.

$$\Delta E = -567\ J + 89\ J = -478\ J$$

4. State functions

A state function is one that depends on the condition of the system and not on how the system arrived at that condition. An example of a state function is the internal energy of a system. The concept of a state function is a very important one for thermodynamics. An everyday example of state functions you are familiar with is the distance between two cities, such as Boston and New York. The distance is a state function, but the mileage one might drive depends on the path taken. Similarly, the elevation at the top of a mountain is a state function, but the amount of work required to climb the mountain depends on the route (path) one takes to get to the top. Both heat and work are not state functions, but they can become state functions if the path is specified. For example, heat change at constant pressure associated with a transformation of one mole of a specified substance is a state function, the enthalpy.

5.A. Heats of reaction and enthalpy change

It is useful to define several terms for our discussion of enthalpy:

A. Heat of reaction—the heat effect in a chemical reaction. The quantity of heat exchanged between a system and its surroundings when a chemical reaction occurs at constant temperature and pressure.

B. Exothermic reaction—in a non-isolated system heat is given off to the surroundings.

C. Endothermic reaction—in a nonisolated system heat is absorbed from the surroundings.

5.B. Enthalpy change and pressure-volume work.

Pressure-volume work is work done as a result of a volume change in the system. If the system expands, the system does work on the surroundings and the energy of the system decreases. If the system contracts, the surroundings do work on the system and the energy of the system increases. Pressure-volume work is $-P\Delta V$ and the sign will be negative for expansion of the system and positive for contraction of the system. If we remember that the ΔV term is $V_{final} - V_{initial}$, then for expansion the final volume is larger, the change in volume term is positive, and $-P\Delta V$ will be negative because of the negative sign. Similarly for contraction, the change in volume is negative and taken with the negative sign, the overall term will be positive. The formula for the enthalpy change is

$$\Delta H = \Delta E + P\Delta V$$

at constant pressure,

$$\Delta H = q_P$$

5.C. Properties of enthalpy.

Enthalpy is an extensive property. The enthalpy change depends upon the particular states of, and the amounts of, all substances undergoing the change. Enthalpy is a state function. It does not depend on the path the system took to arrive at the state.

Enthalpy changes have unique values.

Since enthalpy is a state function, the state of the system must be defined. The enthalpy change is related to the amount of material undergoing the change.

Exercise (text) 6.3A. Use the equation $3\ O_2(g) \dashrightarrow 2\ O_3(g)$ $\Delta H = +285.4$ kJ to calculate ΔH for the reaction $3/2\ O_2(g) \dashrightarrow O_3(g)$

Note: Enthalpy changes are for one mole of substance undergoing the change unless otherwise indicated.

Solution: The enthalpy of the reaction depends on the amount of material. Since the first equation is divided by 2, ΔH for the reaction must also be divided by 2 and therefore ΔH for the second reaction is +142.7 kJ.

Exercise (text) 6.3B. Given the equation

$$2\ Ag_2S(s) + 2\ H_2O(l) \dashrightarrow 4\ Ag(s) + 2\ H_2S(g) + O_2(g) \qquad \Delta H = +595.5\text{ kJ}$$

calculate ΔH for the reaction

$$Ag(s) + 1/2\ H_2S(g) + 1/4\ O_2(g) \dashrightarrow 1/2\ Ag_2S(s) + 1/2\ H_2O(l)$$

Solution: Equation (2) is the reverse of equation (1). Further, equation (2) represents 1/4 the amount of equation (1). Since ΔH must be zero for a cycle (return to the initial state), the enthalpy change for the reverse of a reaction is the same as the forward process, but the sign is reversed.

The reverse of (1) gives:

$4\ Ag(s) + 2\ H_2S(g) + O_2(g) \text{-->} 2\ Ag_2S(s) + 2\ H_2O(l)$ $\Delta H = -595.5$ kJ

The desired reaction is 1/4 of the above reaction. We know that ΔH depends on the amount of material. We divide the equation by 4, that is, divide the stoichiometic coefficients by 4, and also divide ΔH by 4. This gives:

$Ag(s) + 1/2\ H_2S(g) + 1/4\ O_2(g) \text{-->} 1/2\ Ag_2S(s) + 1/2\ H_2O(l)$ $\Delta H = -148.9$ kJ

Therefore ΔH for reaction (2) = {-ΔH for reaction (1)}/4 = -148.9 kJ.

5.D. Enthalpy in stoichiometric calculations

Enthalpy depends on the amount of material in a reaction. We can calculate the quantity of heat involved in much the same way that we relate the number of moles or masses of reactants and products. Exercise 6.3B above illustrates the use to stoichiometry in calculating the heat of reaction.

6. Calorimetry: Measuring quantities of heat

A calorimeter is used to measure quantities of heat. To do this we must remember that whenever heat is lost by a system, it must be gained by the surroundings. To express this, we change the sign of q. For example, an exothermic reaction gives off 100 J of heat. The energy of the system decreases and q for the reaction (system) is -100 J. The surroundings have increased energy and q is +100 J. For energy to be conserved, $q_{sys} = -q_{surr}$. Heat capacity is a measure of the ability of a substance to store energy that is absorbed as heat. The formula for heat capacity is $C = q/\Delta T$, where q is the heat and ΔT is the change in temperature in either C or K.

Exercise (text) 6.6A. Calculate the heat capacity of a sample of brake fluid if the sample must absorb 911 J of heat in order for its a temperature to rise from 15 °C to 100 °C.

Solution: The heat capacity is the heat absorbed per change in temperature. Therefore, C = q/ΔT = 911 J/(100 °C - 15 °C) = 11 J/°C.

7. Molar heat capacity and specific heat.

Since the quantity of heat absorbed by a substance depends on the amount of substance, it is useful to define this dependence in terms of moles and grams. (Note that it takes much longer to boil a large pot of water than to heat a very small pan of water.) Molar heat capacity is the heat capacity for one mole of substance and specific heat is for one gram of substance. Specific heat = heat capacity/mass = q/(m x ΔT). Specific heat will have units of J/g °C and the molar heat capacity will have units of J/mol °C.

Example 6.4. How much heat, in joules, does it take to raise the temperature of 950 g of water from 20 °C to 100 °C? How much heat, in joules, does it take to raise the temperature of 100 g of water by the same amount?

Solution: q = specific heat x m x ΔT, rearranging from the definition of specific heat given above.

The specific heat of water is 4.18 J/g °C. q = 4.18 J/g °C x 950 g x (100 °C - 20 °C) = 3.2×10^5 J. For 100 g of water, the heat required would be q = 4.18 J/g °C x 100 g x (100 °C - 20 °C) = 3.3×10^4 J. (The smaller quantity of water takes much less heat to raise the temperature.)

Exercise (text) 6.8B. How many grams of copper can be heated from 22.5 °C to 35.0 °C by a quantity of heat sufficient to raise the temperature of 145 g H_2O from 22.5 °C to 35.0 °C?

(Use Table 6.1 for values of specific heat.)

Solution: This is a two-part problem. We must first find the quantity of heat from the change in temperature of the water. The second part is to find the mass of copper that will absorb this quantity of heat for the temperature change given. In both cases we need the specific heat of the substance. In Table 6.1, water has a specific heat of 4.18 J g^{-1} °C^{-1} and copper has a specific heat of 0.385 J g^{-1} °C^{-1}. To find the amount of heat from the water, q = specific heat x mass x ΔT = 4.18 J g^{-1} °C^{-1} x 145 g x (35.0 °C - 22.5 °C) = 7.58×10^3 J.

To find the mass of copper, we rearrange the equation to give, m = q/(specific heat x ΔT) = 7.58×10^3 J/(0.385 J g^{-1} °C^{-1} x (35.0 °C- 22.5 °C)) = 1.58×10^3 g. Since the specific heat of copper is less than 0.1 that of water, we would expect to have 10 times as much mass.

In measuring heat changes, ΔH = heat absorbed at constant pressure and ΔE = heat absorbed at constant volume, such as measured in a bomb calorimeter; and if $P\Delta V$ is small, then $\Delta H \sim \Delta E$.

Example 6.5. A 0.480 g sample of graphite is burned with an excess of $O_2(g)$ in a bomb calorimeter having a heat capacity of 5.15 kJ/°C. The calorimeter temperature rises from 25.00 °C to 28.05 °C. Calculate ΔH for the reaction

$$C(graphite) + O_2(g) \longrightarrow CO_2(g)$$

Solution: First we calculate $q_{calorim}$ from the heat capacity and the temperature change. $q_{calorim}$ = heat capacity x ΔT = 5.15 kJ/°C x (28.05 °C - 25.00 °C) = 15.7 kJ

Since the energy of the universe is constant, heat gained or lost by the reaction is equal to the heat lost or gained from the surroundings. In this case, the surroundings are the calorimeter. Therefore

$$q_{rxn} = -q_{calorim} = -15.7 \text{ kJ}$$

To calculate ΔH for the reaction, we need the heat for one mole of graphite,

$\Delta H = 12.01$ g/mol x (-15.7 kJ/0.480g) = -393 kJ/mol

Example 6.6. A student in a laboratory uses a Styrofoam cup as a calorimeter. She mixes 100.0 mL of 1.00 M HCl and 100.0 mL of 1.00 M NaOH both at 22.0 °C. The final temperature of the solution is 28.8 °C. The density of 0.5 M NaCl is 1.02 g/mL and the specific heat is 4.02 J/g °C. Calculate ΔH for the reaction:

$NaOH(aq) + HCl(aq) \dashrightarrow NaCl(aq) + H_2O(l)$

Solution: Since ΔH depends on the quantity of material, we can first calculate the quantity of NaCl that produced the temperature change. Next we use the formula q = Specific Heat x m x ΔT to calculate the heat. ΔH is then q/moles of NaCl.

Calculating the mass of NaCl solution we use density x volume:

m = d x V = 1.02 g/mL x 200.0 mL = 204.0 g

Substituting to get q, the heat:

q = Specific heat x m x ΔT = 4.02 J/g °C x 204.0 g x (28.8 °C - 22.0 °C)

q = 5.58 kJ

Since heat is given off (the temperature of the solution increases), the sign of q_{rxn} must be negative.

$q_{rxn} = -5.58$ kJ.

$\Delta H = q_{rxn}$/moles of NaCl

The amount of NaCl is calculated from the final solution, which is 0.500 M and has a volume of 200.0 mL. The amount is then 0.5 moles/L x 0.2000 L = 0.100 moles.

$\Delta H = q_{rxn}$/moles of NaCl = -5.58 kJ/0.100 moles = -55.8 kJ/mol

For reactions involving gases, we ordinarily use a device called a bomb calorimeter which is a container with strong steel walls where reactants and products are confined to a constant volume.

8. Hess's Law

Hess's law states that the enthalpy change for an overall or net reaction is the sum of enthalpy changes for individual steps in the reaction or process, each corrected for the relative number of moles of substances participating in the overall change.

Example 6.7. Calculate ΔH for the reaction:

$NO(g) + O(g) \rightarrow NO_2\ (g)$

Use the following information:

$NO(g) + O_3(g) \rightarrow NO_2(g) + O_2(g)$ $\Delta H = -200$ kJ

$O_3(g) \rightarrow 3/2\ O_2(g)$ $\Delta H = -143$ kJ

$O_2(g) \rightarrow 2\ O\ (g)$ $\Delta H = 498$ kJ

Looking at the above equations, we see that using the first equation as written will give us the required NO(g). Further this equation uses one mole of NO, as required for the desired reaction. Next we need to obtain one mole of O(g). Reversing the third equation and dividing by 2 will give us the second reactant. Summing these two equations, we now have:

$$NO(g) + O_3(g) + O\ (g) \rightarrow NO_2(g) + 3/2\ O_2(g)$$

In order to remove the O_3 (g) from the reactant side and the 3/2 O_2 (g) from the product, we subtract reaction two. This now gives us the desired reaction. Keeping track of the ΔH values in the same way as the reactions, we have $\Delta H = \Delta H(1) - 1/2\ \Delta H(3) - \Delta H(2)$. This gives the value for the desired reactions as: $\Delta H = (-200\text{ kJ}) - 1/2\ (498\text{ kJ}) - (-143\text{ kJ}) = -306$ kJ. Remember, we must reverse the sign of ΔH if we reverse the reaction and also remember to keep track of the sign as we do the algebra.

9. Standard enthalpy of formation

The enthalpy of formation is the enthalpy change that accompanies the formation of a compound from the most stable forms of its elements in their standard states. Use ΔH_f° to determine the enthalpy changes for chemical reactions.

Exercise (text) 6.15A. Ethylene, derived from petroleum, is used to make ethanol for use as a fuel or solvent. The reaction is

$$C_2H_4(g) + H_2O(l) \text{ --> } CH_3CH_2OH(l)$$

Use data from Table 6.2 to calculate ΔH° for this reaction.

Solution: To calculate ΔH° for the reaction, use the $\Sigma\ \Delta H^\circ_f$(products) - $\Sigma\ \Delta H^\circ_f$(reactants). Using Table 6.2,

$$\Delta H^\circ = (-277.7\text{ kJ}) - \{(52.26\text{ kJ}) + (-285.8\text{ kJ})\} = -44.2\text{ kJ}.$$

10. Ionic Reactions in solution

In order to determine the enthalpy of reactions involving ions, we apply the same techniques as described above. However, because we cannot form only anions or only cations, but we must form both, we arbitrarily assign H^+ (aq) ions an enthalpy of formation of zero.

Skills Test Problem

A runner produces 300.0 mL of sweat. The sweat evaporates at 25 ^{0}C to cool the runner.

a. How much heat does the runner lose by evaporation the sweat? (Assume the sweat is essentially water.)

b. How many Cal (food calories) does the runner burn to maintain her body temperature if all the heat lost by the evaporation of sweat comes from metabolism?

Solution

The reaction for the evaporation of sweat is

$$H_2O\ (l) \rightarrow H_2O\ (g)$$

$$\Delta H = \Delta H^0_{react} - \Delta H^0_{prod} = -241.8\ kJ/mol - (-285.8\ kJ/mol)$$

$$= +44.0\ kJ/mol$$

300.0 mL of water can be converted to moles

300.0 mL x 1 g/mL x 1 mol/18.0 g = 16.7 mol

ΔH for the evaporation is 44.0 kJ/mol x 16.7 mol = 735 kJ

To calculate the number of calories, we assume all of the heat lost must be replaced by metabolism. Remember that one food calorie (Cal) is one kcal.

735 kJ x 1 Cal/4.18 kJ = 176 Cal

Quiz A

1. The standard state of a substance is the
   a. most stable form at 25 °C
   b. the liquid state
   c. pure form at 1 atm
   d. none of these
2. 10.0 g of water at 25 °C is mixed with 120.0 g of water at 100 °C. What is the final temperature of the water?
3. Write the formation reaction for $H_2O(l)$.
4. What is the specific heat of a 20.0 g solution whose temperature is raised 15 °C by the addition of 1023 J of heat?
5. Use Table 6.2 to determine ΔH° for the reaction $CO_2(g) + H_2(g) \text{-->} H_2O(l) + CO(g)$.
6. A calorie is
   a. a quantity of heat
   b. a unit of light
   c. 1000 joules
   d. a vitamin in foods

e. none of the above

7. An endothermic reaction

   a. absorbs heat
   b. has a positive enthalpy change
   c. lowers the temperature in an isolated system
   d. all of the above

8. Given that

   $2\ Cu(s) + O_2(g) \text{-->} 2\ CuO(s) \qquad \Delta H° = -314.6\ kJ$

   What is ΔH° for the reaction

   $CuO(s) \text{-->} Cu(s) + 1/2\ O_2(g)$

Quiz B

1. What is ΔH° for the reaction C(graphite) + 2 S(rhombic) --> $CS_2(l)$? The following reactions can be used:

   $C(graphite) + O_2(g) \text{-->} CO_2(g) \qquad \Delta H° = -393.5\ kJ/mol$

   $S(rhombic) + O_2(g) \text{-->} SO_2(g) \qquad \Delta H° = -296.9\ kJ/mol$

   $CS_2(l) + 3O_2(g) \text{-->} CO_2(g) + 2SO_2(g) \qquad \Delta H° = -1075.2\ kJ/mol$

2. The molar heat of combustion of acetylene (26 g/mol) is -1301 kJ/mol. Combustion of 0.56 g of acetylene produces how many kJ of heat?
3. How much heat is needed to increase the temperature of 15.0 g of water from 15 °C to 100 °C?
4. Define the term exothermic reaction.
5. A system absorbs 250 J of heat and does 430 J of work. What is the change in internal energy of the system?
6. Write the reaction for the formation of $CH_4(g)$.
7. Heat can be defined as

   a. energy transferred as a result of a temperature difference
   b. a measure of the disorder of a system
   c. a characteristic of electrons that gives rise to magnetic properties of an atom
   d. none of these

8. What is the change in internal energy of a system that has 425 J of work done on it and gives off 250 J of heat?
9. What is the device used to measure a quantity of heat?

Quiz C

Fill in the blanks

1. _____________ is the energy transferred from an object to another as the result of a temperature difference between them.
2. Chemical energy is a form of ____________________ where the chemical bonds act as the storage medium.
3. The First Law of Thermodynamics is also know as _________________________.
4. Energy entering a system carries a ____________ sign.
5. ____________________ is the capacity to do work or supply heat.

Problems

6. From Table 6.2 calculate the heat of reaction from the heat of formation for the following reaction:
   $SO_3(g) + H_2O\ (l) \rightarrow H_2SO_4\ (l)$
7. Calculate the heat capacity of a piece or iron metal if 1200 J of heat produces a temperature increase from 100 ^{0}C to 130 ^{0}C.
8. What is the change in internal energy of a system if it absorbs 250 J of heat and does 300 J of work?
9. Write the correct equation to define the enthalpy change that defines the standard enthalpy of formation for C_2H_5OH (l).

Self Test

Matching

1. State function
2. Hess's Law
3. Heat capacity
4. Enthalpy
5. System
6. Work

a. name given to the quantity E + PV
b. everything we focus on in an experiment
c. the amount of heat required to raise the temperature of a substance a given amount
d. the distance moved times the force that opposes the work
e. the overall enthalpy change for a reaction is equal to the sum of the enthalpy changes for the individual steps in the reaction, each corrected for the relative number of moles of substances participating in the overall change.
f. a function or property whose value depends only on the present state of the system, not on the path used to arrive at the condition.

7. Given the reaction
   $H_2(g) + Cl_2(g) \rightarrow 2HCl\ (g)$ $\Delta H = -184.6$ kJ
   What is the molar enthalpy of formation of HCl (g)?

8. Given the reaction

   $2H_2\,(g) + O_2(g) \rightarrow 2H_2O\,(g)$ $\Delta H = -571.6$ kJ

   How much heat is produced from the reaction of 24.9 g of H_2 (g)?

9. Using the reaction is Problem 8, determine how much oxygen must react to produce 500 kJ of heat?

10. The specific heat of mercury is 0.139 J $g^{-1}\ {}^{0}C^{-1}$. How much heat is required to raise the temperature of 4.0 g Hg from 15 ^{0}C to 45 ^{0}C?

11. Use the following reactions to write the balanced equation for the production of hydrogen peroxide and nitrogen gas from hydrazine (N_2H_4) and O_2(g) and calculate ΔH for the reaction.

    $N_2H_4\,(l) + O_2\,(g) \rightarrow N_2\,(g) + 2\,H_2O\,(l)$ $\Delta H = -621.6$ kJ

    $H_2\,(g) + 1/2\,O_2\,(g) \rightarrow H_2O\,(l)$ $\Delta H = -285.8$ kJ

    $H_2\,(g) + O_2\,(g) \rightarrow H_2O_2\,(l)$ $\Delta H = -187.8$ kJ

CHAPTER 7

Atomic Structure

This chapter first describes the experimental evidence for the picture of an atom with emphasis on the discovery of the electron. The second part of the chapter describes some of the properties of light and these ideas are used in the third part to explain the modern view of atomic structure.

Chapter Objectives: After completing this chapter you should be able to:

1. Describe how a cathode ray tube works.
2. Describe the results of Thomson's experiments with cathode ray tubes.
3. State the important property determined by Millikan's experiments.
4. State Thomson's "raisin pudding" model of the atom.
5. Describe Rutherford's experiments and his conclusions.
6. Describe the properties of protons and neutrons and how they were discovered.
7. Know the function of a mass spectrometer.
8. Know the characteristics of waves.
9. Define an electromagnetic spectrum and know the order, in terms of wavelength range, of various types of electromagnetic radiation.
10. Define continuous and line spectra.
11. State the essential points of Planck's quantum theory.
12. Describe the photoelectric effect and calculate the energy of photons.
13. Determine the energy of an electron in the H atom.
14. Calculate the differences in energy levels of electrons in the hydrogen atom.
15. Sketch the line spectrum of hydrogen.
16. Define ground state and excited state of an atom.
17. Describe the wave properties of matter.
18. Briefly define wave mechanics and wave functions.
19. Know the designations of the four quantum numbers and their possible values.
20. Describe atomic orbitals and the shapes of s, p and d orbitals.
21. State the allowed values of the electron spin quantum number, m_s.

Chapter Summary

1. Cathode ray tubes

Cathode ray tubes are evacuated tubes with an anode at one end and a cathode at the other. When high voltage is applied, radiation is emitted by the cathode. These rays (cathode rays) are detected by light emitted when they strike a zinc sulfide coated screen.

Note that negative ions, anions, are attracted to the positive electrode, the anode. Similarly, positive ions, cations, are attracted to the negative electrode, the cathode.

2. Thomson's experiments with cathode rays

Thomson established the ratio of mass (m) to charge (e) for cathode rays. He concluded that cathode rays are negatively charged fundamental particles of matter (electrons) found in all atoms.

3. Electron charge: Millikan's Oil-Drop Experiment

Millikan established the charge (e) of an electron. This information, combined with the m/e ratio determined by Thomson, allowed the mass of the electron to be calculated.

4. Thomson's "Raisin Pudding" Model

Thomson described a model of the atom in which positive charge is uniformly distributed in a sphere and the electrons are imbedded in the sphere. For the hydrogen atom, this model had the electron as the center of the sphere surrounded by a positive charge.

5. Rutherford's Experiments

Unstable nuclei decay by giving off radiant energy. The three principal types of radioactive decay are alpha particles, beta particles and gamma rays. Alpha particles are extremely fast-moving, excited, doubly ionized He nucleus, He^{2+}; beta particles are fast-moving, highly excited electrons; and gamma rays are a form of electromagnetic wave radiation, with wavelengths on the order of the size of nuclei, and are a form of energy and not a form of matter. Matter, however, especially the small particles mentioned above, has dual particle-wave behavior.

By bombarding a thin gold foil with alpha particles, Rutherford determined that most of the particles would pass through the foil, but that some were sharply deflected. From these experiments, Rutherford concluded that the atom was mostly empty space with a center of positive charge concentrated in the nucleus. These experimental results were very startling. Imagine throwing a baseball at a wall. Some of the baseballs pass through the wall without being deflected!

6. Properties of protons and neutrons and how they were discovered

Protons were discovered from experiments with cathode ray tubes. It was found that positive ions were formed as well as electrons. When hydrogen gas was used, the lightest particles (protons) were formed.

A proton has a mass about 1836 times that of an electron. The charge is equal in magnitude, but opposite in sign, to that of an electron. The proton is the fundamental unit of positive charge in all nuclei.

Bombardment of beryllium "target" plate with alpha particles produced a highly penetrating form of radiation that was not deflected by electric or magnetic fields. This experiment, together with earlier experiments that could not account for all the mass in the nucleus, led Chadwick to propose these particles were neutrons.

Neutrons have the same mass as protons, are neutrally charged, and are located in the nucleus.

7. Function of a mass spectrometer

A mass spectrometer separates positive ions according to their mass-to-charge ratios. The ions are separated by deflection in an electric and magnetic field.

A nuclide is an atomic species having a particular atomic number and mass number. The atomic masses and fractional abundancies are determined by mass spectrometry. The weighted average atomic weight (or atomic weight, as it is usually called) is calculated from these data.

Example 7.1. Carbon exists as two isotopes: carbon-12, with a fractional abundance of 0.9890 and a mass of 12.00 u; and carbon-13 with a fractional abundance of 0.0110 and a mass of 13.00335 u. Calculate the atomic weight of carbon.

Solution: We use the contribution from each species to calculate the atomic weight.

mass of nuclide x fractional abundance = contribution

(12.00 u x 0.9890) + (13.00335 u x 0.0110) = 12.01 u

8. The wave nature or light

Wavelength is the distance between any two identical points in consecutive cycles in a wave. The metric unit is usually the nanometer, although frequently we use other units to avoid exponential notation. Frequency is the number of cycles of the wave that pass through a point in a unit of time. SI unit of frequency is the hertz (Hz); 1 Hz = 1 s^{-1}. If two waves are traveling at the same speed, a long wavelength will have a lower frequency. c = distance traveled per unit time = $\lambda\nu$. The speed of light, c, is constant and is 3.00×10^8 m s^{-1}.

Exercise (text) 7.1A. Calculate the frequency, in Hz, of a microwave that has a wavelength of 1.07 mm.

Solution: $\nu = c/\lambda$. We must convert mm to m to agree with the units on the speed of light. λ = 1.07 mm x 1 m/1000 mm = 1.07×10^{-3} m.

$\nu = (3.00 \times 10^8 \text{ m s}^{-1})/(1.07 \times 10^{-3} \text{ m}) = 2.80 \times 10^{11}$ Hz

Exercise (text) 7.1B. Calculate the wavelength, in nanometers, of infrared radiation that has a frequency of 9.76×10^{14} Hz.

Solution: Using the equation c = $\lambda\nu$ and solving for λ, $\lambda = c/\nu$. Substituting in gives: λ = $(3.00 \times 10^8 \text{ m s}^{-1})/(9.76 \times 10^{14} \text{ Hz} \times 1 \text{ s}^{-1}/\text{Hz}) = 3.07 \times 10^{-7}$ m

We often convert Hz to s^{-1} "in our heads" without explicitly writing the conversion factor. Converting to nm gives 3.07×10^{-7} m x 10^9 nm/m = 307 nm.

9. The electromagnetic spectrum

An electromagnetic spectrum is a range of wavelengths and frequencies of electromagnetic radiation. The approximate order (there is some overlap) of the various types of electromagnetic radiation, from shortest to longest wavelength, is:

gamma rays < x-rays < ultraviolet radiation < visible < infrared radiation < microwave < radio waves.

10. Continuous and line spectra

A continuous spectrum is the spreading out of the various components of light into an unbroken spectrum of colors such as a rainbow.

Light emitted by de-excitation of an excited element produces a pattern of lines called a line spectrum. The line spectrum is characteristic of a particular element and can be used to identify it. Each line corresponds to electromagnetic radiation of a specific frequency and wavelength.

11. The essential points of Planck's quantum theory

Can you think of everyday examples of situations where things are quantized? (Produce sold only in unit amounts and change machines might be two examples.)

Planck's Quantum Theory postulates that energy can be gained or lost only in whole number multiples of the quantity $h\nu$, where h is Planck's constant. This theory accounted for observations of radiation emitted by solid bodies heated to incandescence ("black body radiation").

12. The photoelectric effect

The photoelectric effect is the ability of some materials to emit electrons from their surfaces when struck by electromagnetic radiation of high enough frequency. Light below the minimum frequency, the threshold frequency, will not cause emission of electrons no matter how bright the light. The energy of a photon = $h\nu$, where $h = 6.63 \times 10^{-34}$ J s and ν is the frequency. The photoelectric effect can be explained only by using the concept of quantized energy.

Exercise (text) 7.3A. Calculate the energy, in joules per photon, of microwave radiation that has a frequency of $2.89 \times 10^{10}\ s^{-1}$.

$$E = h\nu = 6.626 \times 10^{-34} \text{ J s/photon} \times 2.89 \times 10^{10}\ \text{s}^{-1}$$

$$E = 1.91 \times 10^{-23} \text{ J/photon}$$

Example 7.2. Calculate the wavelength emitted in the far ultraviolet that corresponds to an energy of 1609 kJ/mol photons.

Solution: We can combine the two relevant equations:

$$E = h\nu = \frac{hc}{\lambda}$$

This will give the energy per photon. By multiplying by Avogadro's number, N_A, we can obtain the energy per mole of photons, which is the quantity we are given:

$$E_{photon} \times N_A = E_{mol} = 1609 \text{ kJ/mol}$$

Substituting for E_{photon}:

$$\frac{hc}{\lambda} \times N_A = E_{mol}$$

$$hc \times N_A = E_{mol} \times \lambda$$

Then rearranging gives the wavelength for the photon:

$$\lambda = \frac{6.626 \times 10^{-34}\ J \cdot s \times 3.00 \times 10^{8}\ m\ s^{-1}}{1609\ kJ\ mol^{-1} \times \dfrac{10^{3}\ J}{kJ}} \times \frac{6.02 \times 10^{23}\ photons}{1\ mol}$$

$$\lambda = 0.07437 \times 10^{-6}\ m = 74.37\ nm$$

13. Bohr's Hydrogen atom

Bohr calculated the properties of the hydrogen atom and described the electron energy as having only certain allowable values.

$$E_n = \frac{-B}{n^2}$$

Exercise (text) 7.5A. Calculate the energy of an electron in the level n = 6 of a hydrogen atom.

Solution: Using the above equation, and $B = 2.179 \times 10^{-18}$ J,

$$E_6 = -2.179 \times 10^{-18}\ J/(6)^2 = -6.053 \times 10^{-20}\ J$$

14. Explanation of line spectra

As an electron drops from a higher to a lower energy level, energy is emitted as a photon of light. Because electrons can only have discrete values of energy, the transitions in which energy is emitted produces a line spectrum.

Calculate the differences in energy levels of electrons in the hydrogen atom.

The difference between energy levels can be calculated from the above equation where

$$\Delta E = B\left(\frac{1}{n_i^2} - \frac{1}{n_f^2}\right).$$

Exercise (text) 7.6A. Calculate the energy change that occurs when an electron is raised from the n = 2 to the n = 4 energy level of a hydrogen atom.

Solution: Use the above equation.

$$\Delta E = B\left(\frac{1}{n_i^2} - \frac{1}{n_f^2}\right) = 2.179 \times 10^{-18}\ J\left(\frac{1}{2^2} - \frac{1}{4^2}\right) = 4.086 \times 10^{-19}\ J$$

Exercise (text) 7.7B. Calculate the wavelength, in nanometers, that corresponds to the radiation released by the electron energy-level change from $n_i = 5$ to $n_f = 2$ in a hydrogen atom.

Solution: Since the electron is going to a lower energy level, energy will be given up. The energy given up is calculated as above for the transition. From the relationship $E = hc/\lambda$ we can calculate the wavelength.

$$\Delta E = B\left(\frac{1}{n_i^2} - \frac{1}{n_f^2}\right) = 2.179 \times 10^{-18}\ J\left(\frac{1}{5^2} - \frac{1}{2^2}\right) = -4.576 \times 10^{-19}\ J$$

The minus sign indicates that energy is given up. The energy given up will be the energy of the photon.

$\lambda = hc/E$

$= (6.626 \times 10^{-34}$ J s/photon x 3.00×10^{8} m/s)/(4.576×10^{-19} J)

$= 4.34 \times 10^{-7}$ m x 10^{9} nm/m = 434 nm

15. Sketching the line spectrum of hydrogen

A discontinuous or line spectrum of hydrogen is shown in Fig. 7.20 (text). The Balmer series is the line spectrum resulting from transitions to the n = 2 level from higher energy levels and these spectral lines occur in the visible region.

Eample 7.3. Without doing a detailed calculation, in what region of the spectrum would you expect to see a line spectrum for transitions to the n = 1 level from higher energy levels in the hydrogen atom?

Solution. The change in energy would be larger, more energy given up, than for transitions to the n = 2 level. The $1/n^2$ term is larger. This makes sense because the transition between energy levels is larger. The wavelength is inversely proportional to the energy, $\lambda = hc/E$, and λ therefore would be smaller. The transitions would occur in the ultraviolet region.

16. Ground state and excited state of an atom

An atom is in the ground state when the electrons in an atom are in the lowest energy levels. This is usually the most stable state for an atom.

An atom is in an excited state when electrons in an atom are promoted to higher energy levels. An atom in an excited state eventually emits photons and the excited electron drops back to the lower energy levels, eventually ending up at the ground state.

17. The wave properties of matter

Matter has dual particle and wave properties. A particle with a mass m moving at a speed u is equivalent to a wave with wavelength

$\lambda = h/mu$

This is De Broglie's equation, sometimes called the De Broglie wavelength.

Exercise (text) 7.9A. Calculate the wavelength, in nanometers, of a proton moving at a speed of 3.79×10^3 m s^{-1}. The mass of a proton is 1.67×10^{-27} kg.

Solution: The wavelength is calcu[illegible]d from

$$\lambda = \frac{h}{mu} = \frac{6.626 \times 10^{34} \text{ kg m}^2 \text{ s}^{-1}}{1.67 \times 10^{-27} \text{ kg} \times 3.79 \times 10^3 \text{ m s}^{-1}}$$

$$= 1.05 \times 10^{-10} \text{ m}$$

$$= 0.105 \text{ nm}$$

18. Briefly define wave mechanics and wave functions.

Wave mechanics is a mathematical description of atomic structure based on the concept of wave-particle duality, the Heisenberg uncertainty principle, and the wave nature of electrons. Wave functions are mathematical solutions to the equations of wave mechanics.

19. Quantum numbers and atomic orbitals

Know the designations of the four quantum numbers and their possible values.

1. The principal quantum number (n) designates the main energy level. n is a positive integer (n = 1, 2, 3....).
2. The angular momentum quantum number (l) designates the subshells. l has values of 0 to n - 1 (l = 0, 1, ...n-1). The values of l have orbital designations as follows:

l =	0	1	2	3
Orbital =	s	p	d	f

3. The magnetic quantum number (m_l) designates a particular orbital within a sublevel. m_l has values from $-l$ to $+l$.

Exercise (text) 7.10. Consider the limitations on values for the various quantum numbers, and state whether an electron can have each of the following sets. If a set is not possible, state why not.

a. n = 2, l = 1, m_l = -2

b. n = 3, l = 2, m_l = +2

c. n = 4, l = 3, m_l = +3

d. n = 5, l = 2, m_l = +3

Solution:

a. not allowed. m_l can only have values from -l to +l.

b. allowed

c. allowed

d. not allowed. m_l can only have values from -2 to +2 where l = 2.

Exercise (text) 7.11. Consider the relationship among quantum numbers and orbitals, subshells, and shells to answer the following: (**a**) How many orbitals are there in the 5p subshell? (**b**) What is the total number of orbitals in the principal shell n = 4? (**c**) What subshell of the hydrogen atom consists of a total of seven orbitals?

Solution:

a. For a p subshell, that is, orbital $l = 1$, there are three orbitals regardless of the principal shell.

b. For n = 4, $l = 0,1,2,3$ and $m_l = -3, -2, -1, 0, 1, 2, 3$ for $l = 3$. Similarly for the other l values. For $l = 0$ there is one orbital, for l = 1 there are three orbitals, for $l = 2$ there are five orbitals and for $l = 3$ there are seven orbitals for a total of 1 + 3 + 5 + 7 = 16 orbitals.

c. the $l = 3$ subshell.

20. Atomic orbitals and the shapes of s, p and d orbitals.

Atomic orbitals describe regions of high electron probability. The s orbitals are spherical in shape, the three p orbitals, $l = 1$, are dumbell shaped (Fig. 7.26). The five d orbitals are shown in Fig. 7.27.

21. The allowed values of the electron spin quantum number, m_s

The electron spin quantum number, m_s, has values of +1/2 or -1/2.

Additional Exercise

If an electron and a proton travel at the same speed, which one has the shorter wavelength?

Solution

The wavelength of a particle is given by

$$\lambda = \frac{h}{mv}$$

The wavelength is inversely proportional to the mass. The mass of a proton is about 1840 times greater than an electron. Therefore, the proton has a wavelength about 1840 times less than the electron.

Quiz A

1. Which one of the following types of radiation has the shortest wavelength and the highest frequency?

   a. infrared radiation

   b. red light

   c. ultraviolet radiation

d. green light

2. What is the wavelength of an electron (m = 9.1×10^{-28} g) moving with a velocity of 1/2 the speed of light?

True or False.

3. An excited atom can return to the ground state by emitting electromagnetic radiation.
4. An electron in the n=3 state can go to the n=1 state by absorbing energy.
5. The wavelength and frequency of electromagnetic radiation are inversely proportional to each other.

Problems

6. What is the minimum energy required to excite an electron in the hydrogen atom from the n=1 to the n=3 energy level?
7. When an electron has a principal quantum number of 1 (n = 1), what is the value of the orbital quantum number l ?
8. Radiation has a frequency of 5.0×10^{15} Hz. What is the associated wavelength?
9. What is the energy in joules of a photon of wavelength 2.1×10^{-5} m?

Quiz B

1. What wavelength of light is associated with the n=2 to n=1 transition in the hydrogen atom?
2. Which of the following combinations of quantum numbers would be possible? If a combination is not possible, state why not.

	n	l	m_l	s
a.	3	2	-1	1/2
b.	2	1	0	1
c.	4	4	2	-1/2
d.	1	0	0	1/2

3. How many electrons in an atom can have the quantum numbers n = 3, l = 2?
4. Green light has a wavelength of 540 nm. What is the energy of a photon of green light?
5. An electron (m = 9.1×10^{-28} g) has an associated wavelength of 6.85×10^{-11} m, What is the magnitude of the velocity?
6. The symbol ν is used in spectroscopy to represent
   a. wavelength
   b. frequency
   c. the speed of light
   d. energy
   e. none of these.

7. What is the speed of light, in m/sec, to two significant figures?
8. How many orbitals are there in the 4p subshell?
9. What subshell of the Hf atom consists of a total of five orbitals?

Quiz C

Fill in the blank.

1. ________________ are emitted from the cathode of an evacuated tube when electricity is passed through the tube.
2. ________________ confirmed that cathode rays are negatively charged particles by measuring their deflection in magnetic and electric fields.
3. ________________ determined the charge of the electron.
4. ________________ is an experimental technique used to study beams of positive ions produced by using a perforated cathode.
5. ________________ make up most of the mass of an element.
6. ________________ is the number of cycles of a wave that pass through a point in a unit of time.
7. ____________________ states that energy can only be absorbed or emitted as a quantum or exact multiple of a quantum.

Problems

8. Calculate the energy in joules of a photon that has a wavelength of 530 nm.
9. What is the lowest numbered principal shell in which f orbitals are found?
10. What quantum numbers are associated with the 4p subshell?

Self Test

1. What values can m_l have for a 3d orbital?
2. Calculate the energy required to remove an electron from a hydrogen atom in the ground state.
3. What is the wavelength of infrared light (in nm) with a frequency of 1.28×10^{13} Hz?
4. How long will it take radio frequency waves of 850 kHz to travel from a radio station to a nearby city 200 miles away?
5. What is the wavelength of light produced when an electron in a hydrogen atom undergoes a transition from the n = 3 to n = 2 principal energy levels?
6. Calculate the mass of an electron given that an electron moving at a speed of 2.74×10^6 m/s has a wavelength of 0.265 nm. ($1 \text{ J} = 1 \text{ kg m}^2 \text{ s}^{-2}$).

True or False

7. No two electrons in an atom can have the same set of four quantum numbers.
8. Radiant energy can be described as waves.

9. The longer the wavelength, the slower the speed of light.
10. Frequency and wavelength are directly related.
11. The principal quantum number (n) defines the shape of an orbital.
12. When energy is emitted from an atom an electron goes from a lower energy orbital to a higher energy orbital.

Fill in the Blank

13. The fourth quantum number, m_s, is related to ________________________.
14. The energies available to electrons are ____________________________.
15. Both light and matter are _______________ as well as __________________.
16. ___________________ is a series of discrete lines separated by blank spaces and is characteristic of an element.
17. According to the Einstein relationship, mass and wavelength are ________________ related.

Short Answer

18. Describe the basic postulates of the Bohr model of the hydrogen atom.
19. Describe black-body radiation.
20. What is the photoelectric effect?

CHAPTER 8

Electron Configurations, Atomic Properties, and the Periodic Table

This chapter will describe the electron configurations of atoms, that is, how electrons are arranged in the atoms. The electron configuration is used as a basis for the periodic table and we can see how the periodic table helps us to determine trends and properties among elements.

Chapter Objectives: After completing this chapter, you should be able to:

1. Describe the main differences between orbitals of a hydrogen atom and the equivalent orbitals in multielectron atoms.
2. Know how to write subshell notation and the corresponding orbital diagram.
3. Know the principles that determine electron configurations and how to write the electron configurations.
4. Know how to write electron configurations for elements.
5. Know how to write electron configurations for transition elements.
6. Relate the electron configuration of the elements to their position in the periodic table.
7. Deduce electron configurations by using the four principal blocks of elements in the periodic table: s-block, p-block, d-block, and f-block.
8. Define valence electrons.
9. Know how magnetic properties of an atomic substance relate to the pairing of electrons.
10. Know how the atomic radii vary within a group and across the periodic table.
11. State how the size of ions relates to the parent atom.
12. Define ionization energy and electron affinity.
13. Describe the characteristic properties of metals.
14. Describe the characteristics of the Noble gases.
15. Explain some of the behavior of elements through atomic properties and the periodic table.

Chapter Summary

1. Multielectron atoms

The main differences between the hydrogen atom and multielectron atoms concern the energies of the orbitals. In multielectron atoms, as the number of electrons in the atoms, the number of protons in the nucleus, and the positive charge of the nucleus increase, the attractive force between the nucleus and any given electron increases. This results in a lowering of the electron orbital energy.

A second effect is the screening or shielding of electrons. Electrons in the inner-shell tend to screen or shield electrons in outer orbitals from the full attractive force of the nucleus. For multielectron atoms, there are four points to remember with regard to orbital energies:

1. In a hydrogen atom all subshells of a principal shell are at the same energy level.

2. Orbital energies are lower in multielectron atoms than in the hydrogen atom
3. Subshells of a principal shell are at different energy levels, but all orbitals within a subshell are at the same energy level.
4. In higher-numbered principal shells of a multielectron atom, some subshells of different principal shells have nearly identical energies.

2. Subshell notation and the corresponding orbital diagram

The subshell notation uses numbers to designate a principal shell and the letters s, p, d, or f to identify a subshell. A superscript number following the letter indicates the number of electrons in the designated shell.

Example 8.1. Interpret the electron configurations:

a. $1s^2 2s^2 2p^6 3s^2 3p^6 3d^{10} 4s^2$

b. [↑↓] [↑↓] [↑↓] [↑↓] [↑↓] [↑↓] [↑] [↑] [ ]
1s 2s 2p 3s 3p

Solution:

a. This spdf notation indicates two electrons in the 1s subshell, two in the 2s, six in the 2p, two in the 3s, six in the 3p, ten in the 3d and two in the 4s. This is the electron configuration of zinc (Z = 30). The superscripts indicate the number of electrons in each subshell. By adding them, we obtain the total number of electrons and hence total number of protons in the neutral atom.

b. This orbital diagram indicates two electrons in the 1s subshell, two electrons in the 2s, six electrons in the 2p, two electrons in the 3s and two electrons in the 3p subshell. This is the electron configuration of silicon (Z = 14).

3. Principles that determine electron configurations and writing the electron configurations

The principles can be summarized as:

1. Electrons occupy orbitals of the lowest energy first.
   It may seem obvious that the lowest energy will be the lowest shell, but the order of filling is somewhat more complicated in multielectron atoms, arising from the fact that larger nuclei have a higher charge and more strongly attract electrons and the fact that inner shell electrons will shield electrons in the outer shells. The order of electron filling of orbitals is largely experimentally determined. The observed order of filling subshells is usually 1s, 2s, 2p, 3s, 3p, 4s, 3d, 4p, 5s, 4d, 5p, 6s, 4f, 5d, 6p, 7s, 5f, 6d, 7p.
2. No two electrons can have all four quantum numbers alike (Pauli exclusion principle). Therefore, an atomic orbital can accommodate only two electrons and these electrons must have opposing spins.

3. When filling orbitals, electrons will fill orbitals of identical energy (those within the same subshell) singly and with the same spins before the orbitals are doubly occupied (Hund's rule).

4. Writing electron configurations for elements

To write electron configuration, follow the general procedure given below:

1. Determine the number of electrons in the element. Remember that this will be the same as the atomic number for a neutral element.
2. Place the electrons in the various subshells according to the order of energies given by the periodic table.
3. Do not exceed the maximum number of electrons for any subshell.
4. For orbital diagrams, be sure that the orbitals within a subshell are first filled singly and that the electrons have the same spins before the orbitals are doubly occupied.

Example 8.2. Write out the electron configuration for phosphorus, using the *spdf* notation, the noble-gas-core abbreviated electron configuration, and an orbital diagram.

Solution. Phosphorus is atomic number 15, so it will have 15 electrons. Distributing these into the energy levels with the lowest energy filling first gives:

(Z = 15) P $1s^2 2s^2 2p^6 3s^2 3p^3$

[↑↓] [↑↓] [↑↓] [↑↓] [↑↓] [↑↓] [↑] [↑] [↑]

1s 2s 2p 3s 3p

For the noble-gas-core abbreviated electron configuration, substitute [Ne] for $1s^2 2s^2 2p^6$, so

P $[Ne]3s^2 3p^3$ [Ne] [↑↓] [↑] [↑] [↑]

3s 3p

5. Writing electron configurations for transition elements

Electron configurations for the elements Sc through Zn involve filling of inner shell electrons. The electron configurations are given in Table 8.2 (text). The 4s orbitals have lower energy than the 3d orbitals, so in general, the 4s fills first. However, if you notice, Cr and Cu have only one 4s electron. A $3d^5$ or $3d^{10}$ electron configuration confers some additional stability and there is a tendency to form a half-filled or completely filled d orbital.

Example 8.3. Write the electron configurations of Fe and Cu using *spdf* notation.

Solution: Both of these are transition elements, so we expect the d orbitals to be filling. Fe is atomic number 26. There are 26 electrons in the neutral atom. Using the [Ar] configuration, we can account for 18 electrons ($1s^2 2s^2 2p^6 3s^2 3p^6$). We now have eight electrons to place in orbitals. Since we know that the 4s orbital is lower in energy than the 3d orbitals, the 4s will fill first, leaving 6 electrons for the 3d shell. This gives an electron configuration of $[Ar]3d^6 4s^2$.

Copper is atomic number 29, indicating 29 electrons in the neutral atom. Similar reasoning can be applied to the electron configuration as was used to write the electron configuration of Fe. [Ar] accounts for 18 electrons, leaving eleven. If we now fill the 4s orbital, we have 9 electrons for the 3d subshell. A completely filled subshell is more stable, so the 3d subshell completely fills giving an electron configuration of $[Ar]3d^{10}4s^1$.

6. Relating the electron configuration of the elements to their position in the periodic table.

With only a few exceptions, the number of electrons in the outer shell is the same for each element within a group of the periodic table. For main-group elements, the number of outer shell electrons is the same as the periodic table group number.

Example 8.4. Write out the subshell notation for the outer-shell electrons for gallium and tellurium.

Solution: Gallium is in Group 3A and the 4th period of the periodic table. The Group 3A indicates three outer shell electrons and the 4th period indicates the principal shell is four. The subshell notation is therefore $4s^2 4p^1$. Tellurium is in the 6A Group and the 5th period. Six outer shell electrons will be arranged in the subshell five giving an electron configuration of $5s^2 5p^4$.

7. Electron configurations of the four principal blocks of elements in the periodic table: s-block, p-block, d-block, and f-block.

The s-block are main group elements. The ns subshell fills.
p-block are also main group elements. The np subshell fills.
d-block are transition elements. The (n-1)d subshell fills.
f-block are the actinide and lanthanide series. The (n-2)f subshell fills.

Example 8.5. Refer only to the periodic table and give the complete ground-state electron configuration, in the expanded spdf notation and in the noble-gas-core abbreviated notation, for each of the following:

a. germanium **b.** zinc **c.** titanium **d.** iodine

Solution:

a. Germanium is a p-block element. Since it is in Group 4A, it will have four outer shell electrons. It is also in the 4th period so the subshell will be four. The outer electron configuration will be $4s^2 4p^2$, filling in the lower energy levels to give the complete electron configuration will give: $1s^2 2s^2 2p^6 3s^2 3p^6 3d^{10} 4s^2 4p^2$. The noble-gas-core notation will be $[Ar]3d^{10}4s^2 4p^2$.

b. Zinc is a d-block element. It is in the fourth period which indicates the four subshell will be the outermost shell. The electron configuration will be: $1s^2 2s^2 2p^6 3s^2 3p^6 3d^{10} 4s^2$ and the noble-gas-core designation will be: $[Ar]3d^{10}4s^2$.

c. Titanium is also a transition d-block element. It is also in the 4th period. The 3d level will be the level filling (4s will be filled because it is lower energy). The electron configuration will be: $1s^22s^22p^63s^23p^63d^24s^2$ and the noble-gas-core configuration is $[Ar]3d^24s^2$.

d. Iodine is a main group VIIA element. It has seven outer shell electrons and is in the fifth period. The outer electron configuration is therefore $5s^25p^5$. Filling in the inner levels gives $1s^22s^22p^63s^23p^63d^{10}4s^24p^64d^{10}5s^25p^5$. The noble-gas-core abbreviation is $[Kr]\ 4d^{10}5s^25p^5$.

8. Valence electrons

Valence electrons are the electrons in the outermost occupied principal shell, that is, in the level of highest principal quantum number, n.

9. Magnetic properties of an atomic substance

Paramagnetic atoms or molecules of a substance have unpaired electrons and are attracted to a magnetic field. Diamagnetic atoms or molecules of a substance have all the electrons paired and are weakly repelled by an external magnetic field.

Example 8.6. Which of the following elements, in atomic form, are expected to exhibit paramagnetism?

a. potassium **b.** mercury **c.** barium
d. gallium **e.** sulfur **f.** lead

Solution: In order to determine which elements exhibit paramagnetism, we need to know which elements have unpaired electrons. Those elements with unpaired electrons will be paramagnetic. To determine the unpaired electrons, we need to know the electron configuration.

a. Potassium: the electron configuration is $[Ar]4s^1$. The 4s electron is unpaired and the element is paramagnetic.

b. Mercury: $[Xe]4f^{14}5d^{10}6s^2$ In this case, the f, d, and s orbitals are filled and the electrons paired. Mercury will not be paramagnetic.

c. Barium: This is a main group 2A element so the outer shell electron configuration will be $6s^2$. The electrons are paired and the element is not paramagnetic.

d. Gallium: $[Ar]3d^{10}4s^24p^1$. In this case, one p orbital is half-filled and there is one unpaired electron. The element is paramagnetic.

e. Sulfur: $[Ne]3s^23p^4$. Since there are three p orbitals and four p electrons, two of the electrons will be unpaired. Sulfur is paramagnetic.

f. Lead: $[Xe]4f^{14}5d^{10}6s^26p^2$. Again, there are three p orbitals and two p electrons. Both electrons in the 6p level will be unpaired. Lead will be paramagnetic.

10. Atomic radii vary within a group and across the periodic table.

Atomic radius is the distance from the nucleus to the outer-shell (valence) electrons. Within a vertical group of the periodic table, each element moving down the group has one more principal shell and therefore has a larger atomic radius.

Because the effective nuclear charge for A group elements (1A-8A, excluding B groups elements) tends to increase from left to right in a period of the periodic table, the atomic radii of these elements decrease from left to right in a period.

Exercise (text) 8.6. With reference only to the periodic table, arrange each set of elements in order of increasing atomic radius.

a. Be, F, N **b.** Ba, Be, Ca **c.** Cl, F, S **d.** Ca, K, Mg

Solution: Atomic radius increases from top to bottom within a group and decreases going from left to right in the periodic table.

a. F < N < Be (from left to right, size decreases)

b. Be < Ca < Ba (size increases from top to bottom within a group)

c. F < Cl < S (S is to the left of Cl and, therefore, larger than Cl. F is above Cl in the same group and, therefore, smaller than Cl.)

d. Mg < Ca < K (same logic as in **(c)** above)

11. Ionic radii

Ionic radii are determined by the same principles as atomic radii. Cations (positive ions) are smaller than the atoms from which they are formed. They have lost one outer shell of electrons. Anions (negative ions) are larger than the ions from which they are formed. The gain of outer shell electrons causes increased electron repulsion and a larger atomic radius.

Example 8.7. Arrange the following species in order of increasing radius: Br^-, Rb^+, Se^{2+}, Sr^{2-}, Y^{3+}.

Solution: All of these ions are isoelectronic with electron configurations of $[Ar]3d^{10}4s^24p^6$. Therefore, from the rules given above, we would expect the order of size to be $Y^{3+} < Sr^{2+} < Rb^+ < Br^- < Se^{2-}$.

12. Ionization energy and electron affinity

Ionization energy is the amount of energy required to remove an electron from an atom or ion in the gas phase. Although there are some irregularities, in general, ionization energy decreases from top to bottom a group in the periodic table and increases from left to right. These trends correspond to the trends in size of the atomic radii. As the atomic radii increase, the electrons are further away from the charged nucleus and are less tightly held. Ionization is therefore easier.

Electron affinity is the energy associated with addition of an electron to an atom or ion in the gas phase. There is less correlation between atomic size and electron affinity than for ionization energy; however, the smaller the atom, the more negative is its electron affinity.

Exercise (text) 8.8. Without reference to Fig. 8.15, arrange each set of elements in order of increasing first ionization energy.

a. Be, F, N **b.** Ba, Be, Ca **c.** P, F, S **d.** Ca, K, Mg

Solution:

a. Ionization energy increases from left to right in the periodic table; therefore, Be < N < F.

b. Ionization energy decreases from top to bottom in the periodic table within a group. Therefore, Ba < Ca < Be.

c. From the above two considerations, P < S and S < F, so one would expect the order to be P < S < F. However, the actual order is S<P<F. Removing an electron from a p^3 level in P destroys the space-filling feature of the half-filled level to give a partially-filled p^2 level. This is unfavorable, and P requires more energy.

d. K < Ca since ionization energy increases going from left to right and Ca < Mg, so the order is K < Ca < Mg.

13. Characteristic properties of metals

Generally, metal atoms have small numbers of electrons in their valence shells. Metals form ionic compounds with nonmetals. To do so, they lose valence shell electrons to form cations. The metallic character of elements increases from top to bottom in a group and decreases from left to right in a period of the periodic table. This corresponds to the trend for lower ionization energies. The more metallic elements have lower ionization energies, they lose electrons more easily.

Example 8.8. In each set, indicate which is the more nonmetallic element.

a. O, P **b.** As, S **c.** P, F

Solution: Nonmetallic character decreases from top to bottom in a group and increases from left to right in a period of the periodic table.

a. Since O is above and to the right of P in the periodic table, O is more nonmetallic than P.

b. By the same reasoning S would be more nonmetallic than As.

c. F is more nonmetallic than P.

14. Characteristics of the Noble gases.

The Group 8A elements are called the noble gases. Except for He ($1s^2$), they have outer-shell electron configurations of ns^2np^6, which is a very stable electron configuration. The noble gases are very unreactive and rarely enter into chemical reactions.

15. Explaining some of the behavior of elements through atomic properties and the periodic table

1. Flame colors of elements result when the energy for an electronic transition is emitted in the visible region. The lowest ionization energies occur in the Group 1A metals, and all of them exhibit flame color.

2. Elements that gain electrons easily act as oxidizing agents. The Group 7A elements often act as oxidizing agents.

3. s-Block metals often act as reducing agents and lose electrons.

Quiz A

1. Write the electron configuration for Mg.
2. Which of the following would be expected to be paramagnetic?

   a. Ne   b. Na   c. Ba   d. H^+

3. Which has the larger atomic radius, Na or Mg?
4. Which has the larger ionic radius F^- or Na^+ ?
5. Arrange the following elements in order of increasing first ionization energy. Na, Rb, Cs.
6. A potassium atom must gain or lose how many electrons to achieve a noble gas configuration?
7. Which would you expect to be more nonmetallic, Si or N?
8. What is the general electron configuration of the Noble gas elements?
9. Write the electron configuration for Cl^-.

Quiz B

1. Write the electron configuration for Cr.
2. Arrange the following elements in order of decreasing metallic character (most metallic first).

   Al, Mg, Na.

3. Which of the following elements would you expect to be diamagnetic. Explain why.

   a. Na   b. Ar   c. Al   d. C

4. Which is larger, Ca or Ca^{+2} ?
5. Arrange the following elements in order of increasing first ionization energy.

   S, Se, Ga.

6. Which is larger, Br or Br^- ?
7. A group or family in the periodic table is (a) a horizontal row (b) a vertical column (c) a diagonal element (d) a group of elements that have the same melting point (e) none of these.
8. State Pauli's exclusion principle.
9. Which element has the electron configuration $1s^22s^22p^6$?
10. How many electrons are described in the subshell notation $3p^5$?

Quiz C

1. What is the number of unpaired electrons in a Si atom?
2. Which of the following would have the most metallic character:
   a. Rb b. Ag c. C d. Ne
3. Which of the following would have the largest atomic radii:
   a. He b. Xe c. Ne

True or False

4. The radius of a cation is smaller than the radius of the neutral atom.
5. Hund's rule states that no two electrons in an atom can have the same four quantum numbers.
6. If two or more orbitals with the same energy are available, the electrons are put in orbitals pairwise until all orbitals are filled.
7. Wave functions describe a region of space around the nucleus; an orbital where there is a non-zero probability of finding an electron.
8. The energy levels of various orbitals depends only on the value of *n*.
9. Electrons lost by a metal come from the lowest-energy occupied orbital.
10. Similar electron configurations explain why the elements in a given group have similar chemical properties.
11. The s-block elements consist of the elements in the periodic table in Groups 3A through 8A.

Self Test

Matching

1. Pauli
2. Hund
3. Aufbau principle
4. ground-state configuration
5. ionization energy
6. f-block
7. d-block
8. ionic radius

a. the set of rules that guides the filling of atomic orbitals
b. no two electrons in an atom can have the same four quantum numbers
c. if two or more orbitals with the same energy are available, put one electron in each with parallel spin until all are half full
d. the lowest energy electron configuration
e. a measure of the size of a cation or anion based on the distance between the centers of ions in an ionic compound
f. the actual nuclear charge less the screening effect of other electrons in the atom
g. the energy change that occurs when an electron is added to an atom in the gaseous state
h. the energy required to remove the least tightly bound

	electron from a ground-state atom (or ion) in the gaseous state
9. effective nuclear charge	i. comprises the B Group elements found in the main body of the periodic table
10. electron affinity	j. comprises two series of elements extracted from the main periodic table (lanthanides and actinides) and printed at the bottom of the periodic table

Problems/Short Answer

11. Write the electron configuration for chlorine. How many valence electrons are there?
12. What is the general valence shell electron configuration for the group 5A elements? Which block are these elements located in?
13. Why does Na have a lower first ionization energy than Li?
14. Would you expect the first or second ionization energy of calcium to be larger? Why?
15. Why does the atomic radius decrease in going from left to right across a row of A group elements in the periodic table?
16. Write the outer electron configuration for the noble gas elements.
17. What property or properties of an element determine whether it is a metal?
18. The atomic radius increases with increasing mass within a group. Would you expect the ionization energy to increase or decrease? Why?

CHAPTER 9

Chemical Bonds

Chemical bonds are the forces that hold atoms together in molecules and keep ions in place in solid ionic compounds. In this chapter we will examine some of the properties of materials and how these are determined by the nature of the chemical bonds.

Chapter Objectives:

1. Write the Lewis symbols for elements and compounds.
2. Describe ionic bonds.
3. Know what the octet rule is.
4. Describe the energetics of ionic compound formation.
5. Define a covalent bond and know how to represent a covalent bond in a Lewis structure.
6. Describe how a coordinate covalent bond is formed.
7. Know how elements share electrons to form multiple bonds.
8. Define polar covalent bonds.
9. Define electronegativity.
10. Describe the relationship between electronegativity and bond type.
11. Draw Lewis structures for polyatomic molecules.
12. Practice the strategies and know the basic techniques for writing Lewis structures.
13. Use formal charge to write the most plausible Lewis structure.
14. Know how to write resonance structures for a given molecule or ion.
15. Describe exceptions to the octet rule.
16. Define bond length.
17. Define bond dissociation energy.
18. Use bond energies to estimate enthalpy changes.
19. Know the difference between saturated and unsaturated hydrocarbons.

Chapter Summary

1. Lewis symbols for elements and compounds

Lewis symbols are a shorthand way of writing electron configurations. The inner core of electrons is represented by the symbol for the element. The outer electrons are represented as dots around the symbol.

Example 9.1. Give the Lewis symbols for each of the following atoms.

a. Ar **b.** Ca **c.** Br **d.** As **e.** K **f.** Se

Solution: First determine the number of outer shell electrons. Represent these by dots and place them around the atomic symbol.

a. Ar has eight outer shell electrons:

••
: Ar :
••

b. Ca has two outer shell electrons:

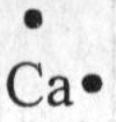

•
Ca•

c. Br has seven outer shell electrons:

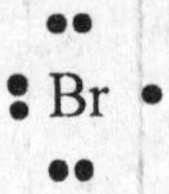

••
: Br •
••

d. As has five outer shell electrons:

••
: Ar :
••

e. K has one outer shell electron:

K•

f. Se has six outer shell electrons:

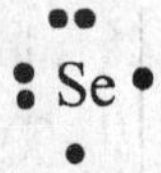

••
: Se •
•

2. Ionic bonds

Ionic bonds are formed between two atoms when one atom loses one or more electrons and the other gains one or more electrons. The resulting ions have opposite charges and are strongly attracted. Each ion tends to form an ideal gas electron configuration.

3. The octet rule

In many cases, elements tend to give up or gain electrons to form a noble gas configuration of eight outer shell electrons. This is known as the octet rule.

Example 9.2. Use Lewis symbols to show the transfer of electrons from aluminum to oxygen atoms to form ions with noble gas configurations. What is the formula and name of the product?

Solution: Al will lose three electrons and oxygen will gain two electrons so that both will have a noble gas configuration.

$2\ Al + 3/2\ O_2 \text{ --> } Al_2O_3$ (The compound is aluminum oxide.)

$2\ Al + 3\ O \text{ --> } 2\ Al^{3+}\ 3\ O^{2-}$

4. Energetics of ionic compound formation

The Born-Haber cycle describes the energetics of ionic compound formation. For NaCl these can be summarized as:

1. Na(s) --> Na(g) (energy of sublimation)

2. $Cl_2(g) \rightarrow 2\ Cl(g)$ (bond dissociation energy)
3. $Na(g) \rightarrow Na^+(g) + e^-$ (first ionization energy)
4. $Cl(g) + e^- \rightarrow Cl^-(g)$ (electron affinity)
5. $Na^+(g) + Cl^-(g) \rightarrow NaCl(s)$

These reactions provide a series of steps, which overall can be summarized as $Na(s) + 1/2\ Cl_2(g) \rightarrow NaCl(s)$.

Exercise (text) 9.3B. Use the data provided in the text and the enthalpy of formation of lithium chloride, $\Delta H_f^\circ = -409$ kJ/mol LiCl(s), to determine the lattice energy of LiCl.

Solution: The equation we want is $Li^+(g) + Cl^-(g) \rightarrow LiCl(s)$, which by definition is the lattice energy.
The equation for the enthalpy of formation by definition is

$Li(s) + 1/2\ Cl_2(g) \rightarrow LiCl(s)$ $\Delta H_f^\circ = -409$ kJ/mol (1)

We want to end up in the overall equation with $Li^+(g)$ and Cl^- (g) as reactants and LiCl(s) as product to obtain our desired reaction.

To get $Li^+(g)$ as a reactant we can use the first ionization energy reaction, reverse it and change the sign of ΔH:

1. $Li^+(g) + e^- \rightarrow Li(g)$ $\Delta H_1 = -520$ kJ (the reverse of the first ionization energy)

To obtain $Cl^-(g)$ we use the electron affinity:

$Cl(g) + e^- \rightarrow Cl^-(g)$ and to obtain $Cl^-(g)$ as a reactant we again reverse the reaction:

2. $Cl^-(g) \rightarrow Cl(g) + e^-$ $\Delta H_2 = 349$ kJ

Summing reactions (1) and (2) gives: $Li^+(g) + Cl^-(g) \rightarrow Li(g) + Cl(g)$. Our lattice energy reaction requires LiCl(s) as a product. We can obtain this from the formation reaction. However, we need to cancel the Li(g) as a product and Li(s) as a reactant. The reaction to use is the sublimation reaction,

$Li(s) \rightarrow Li(g)$ and reverse the reaction (and the sign of ΔH)

$Li(g) \rightarrow Li(s)$ $\Delta H_3 = -161$ kJ

and adding to our summed equation, (1) + (2) gives

$Li^+(g) + Cl^-(g) \rightarrow Li(g) + Cl(g)$

$\underline{Li(g) \rightarrow Li(s)}$

3. $Li^+(g) + Cl^-(g) \rightarrow Li(s) + Cl(g)$

Keeping in mind the formation equation, $Li(s) + 1/2\ Cl_2(g) \rightarrow LiCl(s)$, we are almost finished with converting Li. Let's look at Cl and change Cl(g) as a product to $1/2\ Cl_2(g)$. We can use the bond dissociation energy to obtain $Cl_2(g) \rightarrow 2\ Cl(g)$, reversing this by dividing by two gives

$Cl(g) \rightarrow 1/2\ Cl_2(g)$ $\Delta H_4 = 122$ kJ

Adding this to our summed reaction (3) gives

$Li^+(g) + Cl^-(g) \rightarrow Li(s) + 1/2\ Cl_2(g)$

and now adding the formation reaction,

$Li(s) + 1/2\ Cl_2(g) \rightarrow LiCl(s)$ $\Delta H_f^\circ = -409$ kJ/mol

This gives the desired equation:

$Li^+(g) + Cl^-(g) \rightarrow LiCl(s)$

In summary:

$Li^+(g) + e^- \rightarrow Li(g)$ $\Delta H_1 = -520$ kJ (- the first ionization energy)

$Cl^-(g) \rightarrow Cl(g) + e^-$ $\Delta H_2 = 349$ kJ (- electron affinity)

$Li(g) \rightarrow Li(s)$ $\Delta H_3 = -161$ kJ (- sublimation energy)

$Cl(g) \rightarrow 1/2\ Cl_2(g)$ $\Delta H_4 = -122$ kJ (- 1/2 bond dissociation energy)

$Li(s) + 1/2\ Cl_2(g) \rightarrow LiCl(s)$ $\Delta H_f^\circ = -409$ kJ/mol (formation)

$Li^+(g) + Cl^-(g) \rightarrow LiCl(s)$ (lattice energy)

$\Delta H_{net} = (-520\text{ kJ/mol} + 349\text{ kJ/mol} - 161\text{ kJ/mol} - 122\text{ kJ/mol} - 409\text{ kJ/mol})$

$= -863$ kJ/mol

5. Covalent bonds

A covalent bond results from a shared pair of electrons. In writing Lewis structures, we count the shared pair of electrons is for each atom to determine whether the atom has an octet of electrons.

Example 9.3. Write the Lewis structure for Br_2 and indicate the bonding pair(s).

Solution: Each Br has seven valence electrons, for a total of 14 in the Br_2 molecule. This gives seven electron pairs. Using one pair for a covalent bond (the bonding pair), and distributing the other six pairs gives a complete octet around each atom.

Br : Br

bonding pair

:Br:Br: (with lone pairs above and below each Br)

Lewis structure

6. Formation of a coordinate covalent bond

A coordinate covalent bond is formed when both electrons in the shared pair are provided by one element

7. Multiple bonds

Elements can share more than one pair of electrons resulting in multiple bonds (double bond = 2 pairs, triple bond = 3 pairs).

8. Polar covalent bond

Polar covalent bonds result from the unequal sharing of electrons.

9. Electronegativity

Electronegativity is a measure of the tendency of an atom to attract bonding electrons to itself when it is in a molecule. Within a period, elements become more electronegative from left to right. Within a group, electronegativity decreases from top to bottom.

Exercise (text) 9.4A. Refer only to the periodic table and arrange the following sets of atoms in order of increasing electronegativity.

a. Ba, Be, Ca **b.** Ga, Ge, Se **c.** Cl, S, Te **d.** Bi, P, S

Solution: Keeping in mind the two trends described above; the order will be

a. $Ba < Ca < Be$ **b.** $Ga < Ge < Se$ **c.** $Te < S < Cl$ **d.** $Be < P < S$

10. The relationship between electronegativity and bond type

A bond between two atoms of equal electronegativity is nonpolar. The electrons are shared equally. When the electronegativity difference is large, electron transfer occurs and the bond is ionic. Between these two extremes are polar covalent bonds where the electrons are attracted more strongly to one atom than the other.

11. Lewis structures for polyatomic molecules

Nonmetallic atoms often form covalent bonds equal in number to eight minus the group number.

Example 9.4. Draw the Lewis structure for methanol, CH_3OH.

Solution: C is a group 4A element and will form four bonds. Oxygen will form two bonds. Distributing the electrons around carbon,

```
  •
• C •
  •
```

we use three electrons to share with the three hydrogen atoms and one electron to share with the oxygen. Oxygen uses one electron to form the C-O bond and one to form an OH bond.

```
    H
    ••
H : C : O : H
    ••
    H
```

The remaining four electrons form two unshared pairs. After placing these unshared pairs on the oxygen, C and O each have an octet of electrons and each H has two electrons.

```
     H
     ••  ••
H : C : O : H
     ••  ••
     H
```

12. Strategies for writing Lewis structures

Exercise (text) 9.6B. Write the Lewis structure of ethyl chloride, C_2H_5Cl.

Solution: First write the skeleton structure. Since C is the least electronegative, it will be in the center. H atoms only form one bond and will be terminal atoms:

```
  H   H
  |   |
H-C—C-Cl
  |   |
  H   H
```

Determine the number of electrons in the valence shell for each atom and arrange these electrons to give each H two electrons and the other atoms eight. This gives chlorine three unshared electron pairs.

```
    H  H
    ••  ••  ••
H : C : C : Cl :
    ••  ••  ••
    H  H
```

Example 9.5. Write a plausible Lewis structure for carbonyl sulfide, COS.

Solution: There are 4 + 6 + 6 = 16 valence electrons. C is the least electronegative and will be the central atom.

O-C-S This uses four of the available electrons.

Placing the remaining electrons around the atoms, we can put six around oxygen (three unshared pairs) and six around S.

```
 ••      ••
: O : C : S :
 ••      ••
```

Both O and S have completed octets, but C is left with only four electrons. By forming double bonds between O-C and C-S we can complete the octet for C as well as O and S.

:Ö = C = S̈:

This leaves two unshared electron pairs around S and two unshared pairs around O.

Example 9.6. Write a plausible Lewis structure for the phosphonium ion, PH_4^+.

Solution: The number of electrons is 4 + 5 -1 = 8. There is one less electron because of the positive charge.

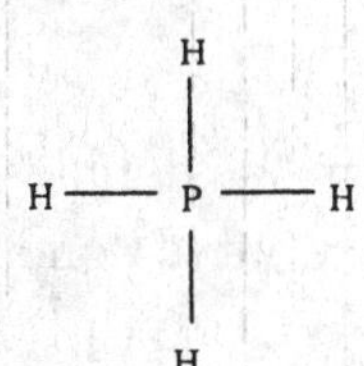

Each H has two electrons and the central P has eight.

Exercise 9.8B (text). Write a plausible structure for the nitronium ion, NO_2^+.

Solution: The number of electrons is 5 + 6 + 6 - 1 = 16. We subtract one for the positive charge. The skeleton structure is

O — N — O

Nitrogen is less electronegative and will be the central atom. Assigning the remaining twelve electrons (four are used in bonds) around the O would give

:Ö — N — Ö:

This leaves the central N with only four electrons. By forming double bonds with oxygen as shown

Ö = N = Ö

all sixteen electrons are assigned and all three atoms have an octet of electrons.

13. Formal Charge

Formal charge is the number of valence electrons in a free neutral atom minus the number of electrons assigned to the atom in a Lewis structure. Know how to assign the electrons. The "best" Lewis structure is usually the one with the lowest formal charge.

Example 9.7. Write the most plausible Lewis structure for NOCl.

Solution: If we write the skeleton structure O-N-Cl we have 6 + 5 + 7 = 18 electrons. Four are used for bonding, so the remaining 14 must be distributed. We could distribute six around O, six around Cl and have one unshared pair on N. This would only give N six electrons. To give N eight, we can form a double bond, but should this be with Cl or O? If we examine the formal charge with a double bond between O and N,

:O = N — Cl: :O — N = Cl:

we have a formal charge of 0 on Cl, 0 on N, and 0 on O. If we form the double bond with Cl, we have a formal charge of +1 on Cl, 0 on N, and -1 on O. The double bond should go between O and N as shown above.

14. Resonance Structures

Remember that resonance structures differ only in the placement of electrons.

Exercise (text) 9.10. Write three equivalent Lewis structures for the nitrate ion, NO_3^-.

Solution: There are (3 x 6) + 5 +1 = 24 electrons. N will be the central atom.

O
|
O — N — O

We can put six unshared electrons around each O, but this leaves N with only six electrons. Therefore we need to form a double bond. The double bond can form with any of the O atoms and therefore we can write six equivalent (resonance) structures:

:O:
||
:O — N — O:

:O:
|
:O — N = O

:O:
|
O = N — O:

15. Exceptions to the octet rule

There are some molecules that do not satisfy the octet rule. These are molecules with an odd number of electrons, molecules with incomplete octets, and molecules with expanded octets.

Example 9.8. Write a Lewis structure for ClF_3.

Solution: There are 4 x 7 = 28 electrons. Write the skeleton structure:

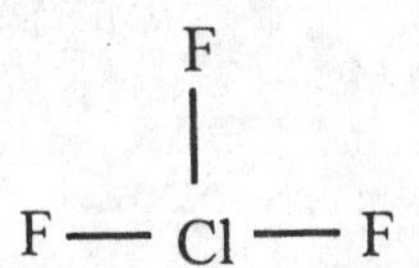

In order to accommodate all the electrons, we must have double bonds between F-Cl, which would give an expanded octet of electrons around both F and Cl or we have the unshared pairs around Cl. An expanded octet on F is forbidden because there are no available orbitals. In this case, Cl has two unshared pairs of electrons.

:F:
:F — Cl — F:

Example 9.9. Only one of the following Lewis structures is correct. Identify that one and indicate the error(s) in the others.

a. Chlorine dioxide :O — Cl — O:

b. Hydrogen peroxide H — O — O — H

c. dinitrogen difluoride :F — N — N — F:

Solution:

a. There are (2 x 6) + 7 = 19 electrons. The structure shown has 20 electrons.

b. This structure has 14 electrons. There are 6 + 6 + 1 +1 = 14 electrons. If we further check the formal charges, they are zero. This is an acceptable structure.

c. Unacceptable; one N has only 6 e^-

16. Bond length

Bond length is the distance between the nuclei of two atoms joined by a covalent bond. How does bond length relate to the strength of the bond? The shorter the bond length, the more tightly two atoms are bound and the stronger the bond.

Example 9.10. How would you expect the indicated bond lengths to compare to single, double or triple bond lengths?

a. The nitrogen to nitrogen bond in CH_3NNCH_3

b. The oxygen to fluorine bond in OF_2

Solution: First draw a plausible Lewis structure.

a.

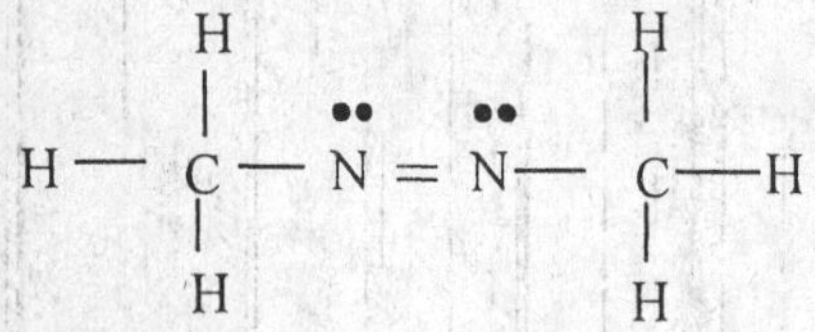

```
      H                 H
      |    ••    ••     |
H —   C — N  =  N —    C—H
      |                 |
      H                 H
```

We would therefore expect an N=N double bond length.

b.

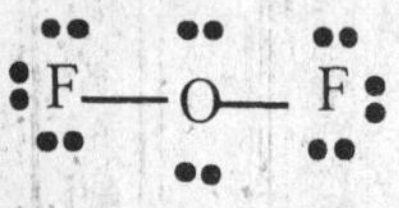

```
 ••    ••    ••
:F — O — F:
 ••    ••    ••
```

This arrangement gives an octet around each atom. The bond length will be a single bond length.

17. Bond dissociation enthalpy

The bond dissociation enthalpy is the enthalpy required (absorbed) to break one mole of covalent bonds between two atoms, with both reactants and products in the gas state.

18. Using bond energies to estimate enthalpy changes

The enthalpy of the bonds broken and the bonds formed are added as in Hess's Law. These estimates are not as accurate as those from enthalpy of formation, but are useful when such data is not available. ΔH for bonds formed will be the negative of the bond energy.

Example 9.11. Estimate the enthalpy of formation for 1.0 mol of $H_2O(g)$ using bond energies.

Solution: First determine the formation reaction, write the Lewis structures and then use the Lewis structures to determine the bonds formed and the bonds broken. Use the bond energies to estimate the enthalpy of formation.

The formation reaction for two moles of H_2O is:

$2\ H_2 + O_2 \text{ --> } 2\ H_2O$

(We are using the reaction for two moles because it is easier to deal with breaking an O_2 bond and less confusing than trying to figure out how to deal with 1/2 O_2 in the reaction.)

Write the Lewis structures:

2 H:H + :Ö=Ö: → 2 H—Ö—H (with lone pairs)

Bonds broken are:

2 H:H = 2 x bond energy for H-H bond = 2 x 436 kJ
1 O::O = 1 x bond energy for O-O double bond = 498 kJ

Sum of energies for bonds broken = 1370 kJ
Bonds formed are:

2 mol H:O for each 1 mol H_2O x 2 mol H_2O formed = 4 x bond energy for H:O
= 4 x (-464 kJ) = -1856 kJ

The enthalpy of reaction is
$\Delta H = \Delta H_{\text{bonds broken}} + \Delta H_{\text{bonds formed}}$ = 1370 kJ - 1856 kJ = -486 kJ

This is for 2 mol of H_2O, however. The enthalpy of formation for 1 mol will be 1/2(-486 kJ) = -243 kJ. The tabulated value of $\Delta H_f^\circ[H_2O(g)]$ is -242 kJ/mol.

Notice that the enthalpy values must include the correct sign, and that the sign of the bond energy for bonds formed will be opposite that of bonds broken.

19. Saturated and unsaturated hydrocarbons

Know the general formulas for alkanes, alkenes, and alkynes. What is meant by delocalized electrons?

Unsaturated hydrocarons have double or triple bonds between carbon atoms. Alkanes are saturated hydrocarbons and have the formula C_nH_{2n+2}. Simple alkenes have one double bond and the formula C_nH_{2n}. Alkynes have carbon to carbon triple bonds.

Delocalization occurs when electrons are not localized between a pair of atoms but can move throughout the molecule. Benzene has delocalized electrons and the bonds between carbon atoms are intermediate in length between single and double bonds.

20. Polymers

A large number of simple molecules (monomers) are joined together into a very large molecule called a polymer. Proteins, made up of amino acids, and DNA, made up of nucleosides, are two examples of biological polymers.

Quiz A

1. Draw the Lewis structure of HClO.
2. Which of the following molecules contains the shortest bond distance?
   a. N_2 b. F_2 c. CO_2 d. CH_4 e. O_2
3. Which of the following exhibits covalent bonding?
   a. NaCl b. ICl c. KF d. SrI_2 e. none of these
4. Given the bond enthalpies C-H (414), H-Cl (431), C-Cl (331), Cl-Cl (243) in kJ/mol, compute $\Delta H°$ for the reaction: $CH_4(g) + 4\ Cl_2(g) \rightarrow CCl_4(g) + 4\ HCl(g)$.
5. Draw the Lewis structure for CO_2.
6. Arrange the following atoms in order of increasing electronegativity: P, O, F.
7. Give a definition of covalent bond.
8. Why does Ar tend not to form chemical bonds?
9. Write the Lewis structure for C_2H_2.

Quiz B

1. Which of the following have ionic bonds?
   a. NH_3 b. H_2O c. $CaBr_2$ d. NOCl e. none of these
2. When a molecule is best represented by more than one electronic structure, it is described in terms of
   a. electronegativity
   b. resonance
   c. multiple bonds
   d. hybridization
   e. none of these
3. Draw the Lewis structure for C_2H_4
4. Arrange the following in order of increasing electronegativity: B, N, F
5. Predict the N-N bond order of N_3^- in terms of single, double, and triple.
6. Give an example of an unsaturated hydrocarbon.
7. Draw the Lewis structure for the simplest compound of C and Br assuming the octet rule is followed.
8. Write the Lewis structure for OH^- ion.
9. Define ionic bond.
10. What is meant by an expanded octet? Give an example.

Quiz C

Fill in the blank

1. ________________ is the amount of energy necessary to break a chemical bond in a isolated molecule in the gaseous state.
2. ________________ represents how an atom's valence electrons are distributed in a molecule.
3. ________________ is the ability of an atom in a molecule to attract the shared electrons in a bond.
4. ________________ are attractive forces between positive and negative ions, holding them together in solid crystals.
5. ________________ is a covalent linkage in which two atoms share two pairs of electrons between them.

True or False

6. A lone pair is a pair of electrons shared between two atoms in a molecule.
7. The formation of a covalent bond leads to lower energy.
8. The bond order refers to the number of electrons in the valence shell.
9. The longer the bond length in a covalent bond, the stronger the bond.
10. Some elements are able to expand their octets because they have unfilled d orbitals.

Self Test

Match the following:

1. alkene — a. electron pairs assigned exclusively to one of the atoms in a Lewis structure.
2. alkyne — b. a measure of the tendency of atoms in molecules to attract electrons to themselves.
3. bond dissociation energy — c. a bond formed by a pair of electrons shared between atoms.
4. bonding pair — d. a pair of electrons shared between two atoms in a molecule.
5. electronegativity — e. a hydrocarbon whose molecules contain at least one carbon-to-carbon double bond
6. covalent bond — f. attractive forces between positive and negative ions holding them together in solid crystals
7. lone pair — g. quantity of energy required to break one mole of bonds of that type in a gaseous species.
8. ionic bond — h. a hydrocarbon whose molecules contain at least one carbon-to-carbon triple bond.

Problems/Short Answer

9. Draw the Lewis structure for AsF_5.

10. Use formal charge to predict the most plausible Lewis structure for BF_3.
11. Which has the most polar bond?

    a. B-C b. B-N c. B-S d. C-C
12. Write the electron configuration of two ions that have noble gas electron configurations.
13. Use bond energies from Table 9.1 to estimate the enthalpy (ΔH) change for the reaction

    $CH_4(g) + 2O_2(g) \rightarrow CO_2(g) + 2H_2O\ (g)$
14. Write the Lewis structure for SO_2.
15. Arrange the following in order of increasing bond strength: single covalent bond; double covalent bond; triple covalent bond.

CHAPTER 10

Bonding Theory and Molecular Structure

In this chapter we will use quantum mechanics to explain some of the properties of chemical bonding and molecular structure beyond the Lewis theory.

Chapter Objectives: You should be able to:

1. Describe molecules in terms of geometrical shapes.
2. Know the difference between electron group geometry and molecular geometry.
3. Know the geometric arrangement of electron groups.
4. Apply VSEPR theory to determine molecular geometry.
5. Predict molecular shapes for molecules with more than one central atom.
6. Define a dipole moment and predict whether a molecule will have a dipole moment.
7. Describe valence bond theory.
8. Describe hybridization of atomic orbitals.
9. Predict hybridization and orbital overlap in a molecule.
10. Predict the hybridization used in forming multiple bonds.
11. Describe geometric isomers and the bonding that leads to geometric isomers.
12. Define molecular orbitals.
13. Describe the bonding in benzene and aromatic compounds.

Chapter Summary

1. Geometric shapes of molecules

Shapes of molecules are described in terms of the positions of the nuclei. Valence-Shell Electron-Pair Repulsion (VSEPR) theory allows us to make predictions about molecular shapes. The basis of these predictions is that valence electrons in bonded atoms repel each other.

2. The difference between electron group geometry and molecular geometry

Electron group geometry describes the arrangement of the electron pairs on the central atom, while molecular geometry describes the shape of the atoms in the molecule.

3. The geometric arrangement of electron groups:

2 electron groups: linear

3 electron groups: trigonal planar

4 electron groups: tetrahedral

5 electron groups: trigonal bipyramidal

6 electron groups: octahedral

Only in cases where there are no unshared electron groups on the central atom will the molecular geometry be the same as the electron geometry.

Exercise (text) 10.2A. Use the VSEPR method to explain the "seesaw" molecular geometry of SF_4 shown in Table 10.1.

Solution: The "seesaw" structure is the result of S having five electron pairs and forming a trigonal bipyramid electron geometry. The unshared pair of electrons occupies one of the positions in the molecule. This unshared pair will occupy a position in the plane of the other two F atoms rather than above or below the plane because the lone pair tends to take up more space than do the bonding pairs. In this planar position, the unshared electrons are 120° from two F atoms and 90° from two F atoms. If the unshared electrons were above or below the plane, the lone pair would be 90° from three F atoms.

4. VSEPR method and molecular geometry

A strategy for applying VSEPR method can be outlined as follows:

1. Write the Lewis structure of the molecule or polyatomic ion.
2. Determine the bonding and lone pairs of electrons around the central atom.
3. Identify the geometric orientation of the electron groups around the central atom (i.e. linear, trigonal planar, etc.)
4. Describe the geometrical shape of the molecule or ion based on the positions around the central atom that are occupied by other atoms (not lone pairs). Use Table 10.1 in the textbook as a guide.

Multiple bonds in VSEPR method are treated the same as single bonds.

Example 10.1. Use the VSEPR method to predict the shape of the ICl_4^- ion.

Solution: First, write a plausible Lewis structure. In this case, the four Cl atoms will be bonded to I with complete octets. Iodine will have an expanded octet including two unshared pairs of electrons.

```
        ..
       :Cl:
        |
  ..   .|.    ..
 :Cl—   I   —Cl:
  ..    |     ..
        |
       :Cl:
        ..
```

Next, determine the electron geometry. I has six electron pairs; two unshared pairs and four bonding pairs. The electron geometry will be octahedral. The molecular geometry will be square planar. The four Cl atoms will be in the same plane with an unshared electron pair above the plane and another below the plane.

5. Molecular shapes for molecules with more than one central atom

For molecules with more than one central atom, work out the geometry around each central atom and combine the results for the overall description of the molecule as far as possible with VSEPR theory.

Example 2. Use the VSEPR method to describe the geometric shape of the ethanol molecule, C_2H_5OH, to the extent that you can with VSEPR theory.

Solution: As before, we start by writing the Lewis structure:

```
    H  H
    ‥  ‥  ‥
H : C : C : O : H
    ‥  ‥  ‥
    H  H
```

For C_1, we have four shared electron pairs and the geometry will be tetrahedral. The same is true for the second C as well. The bond angle between H-C-C will be 109° as will the angle between C-C-O. The oxygen has two bonding pairs of electrons and two unshared pairs. The electron geometry will be tetrahedral and the molecular geometry will be bent with slightly less than 109° bond angles. The C-O-H bond angles will be less than 109°.

Exercise 10.3A (text). Use the VSEPR method to describe the molecular geometry of dimethyl ether $(CH_3)_2O$, as completely as you can.

Solution. Step 1 is to write a plausible Lewis structure

```
    H       H
    ‥   ‥   ‥
H : C : O : C : H
    ‥   ‥   ‥
    H       H
```

This structure has eight electron groups around each atom (except, of course, H). For the central oxygen atom the electron-group geometry will be tetrahedral, as will the geometry of the carbon atoms.

The molecular geometry for each carbon atom will be the same as the electron-group geometry since there are no lone pair electrons. Each carbon will have a tetrahedral geometry. The central oxygen atom has two lone pairs, an AX_2E_2 structure, and will therefore have an angular orientation similar to H_2O molecule, although not necessarily the same bond angles.

6. Dipole moments

The dipole moment of a molecule is a measure of the separation between the centers of the positive and negative charge. A molecule having a dipole moment is said to be polar.

In order to predict whether a molecule has a dipole moment, we must determine the centers of positive and negative charge of the polar bonds. Bond moments, if equal and opposite in direction, will cancel each other out and the molecule will not have a dipole moment.

Exercise (text) 10.4A. Explain whether you expect the following molecules to be polar or nonpolar:

a. BF_3 **b.** SO_2 **c.** BrCl **d.** SO_3^{2-}

Solution:

a. The Lewis structure for BF_3 gives a trigonal planar molecular geometry.

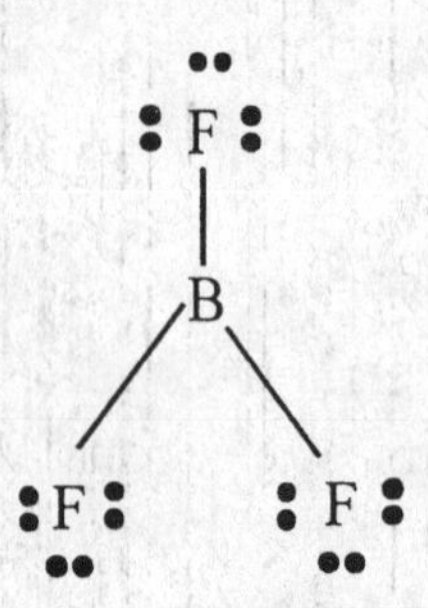

The B will be slightly positive and the F negative; but all B-F dipoles will cancel due to the geometry.

b. The Lewis structure for SO_2 will have 18 electrons or 9 pairs. Two pairs will be used to bond O-S-O. Now place three pairs around one O, two pairs around the other O and form a double bond. One pair will be left for the S. This will give a structure with each atom having an octet. There will actually be two resonance structures.

O= S — O ↔ O—S = O

The S will have an electron geometry of trigonal planar and the molecular geometry will be bent. Because of the bent geometry, the partial negative charges of the O will not cancel out and the molecule will have a dipole moment.

c. BrCl, of course, is a linear molecule as are all diatomic molecules. To determine if it has a dipole moment, we need to decide if there is a difference in electronegativity. From an examination of the periodic table, we see that the electronegativity difference is very small. This molecule would have a very small dipole moment and would be only slightly polar. The Cl is slightly more negative than the Br end of the molecule.

d. The Lewis structure for SO_3^{2-} will be a structure in which S is the central atom and there is a double bond between one of the O atoms (We can actually write three resonance structures.)

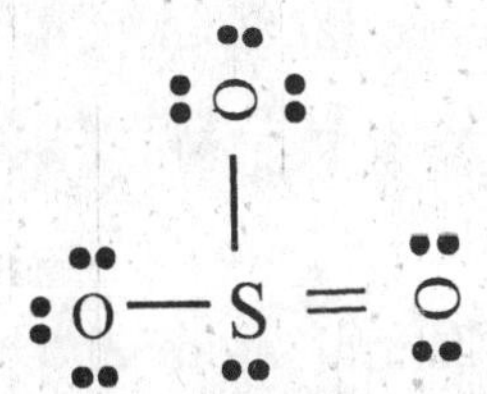

S will have an unshared pair of electrons. This will make the electron geometry tetrahedral and the molecular geometry trigonal pyramidal. The O-S bond will be polar with the partial negative charge towards the O. The molecular geometry will give the molecule a dipole moment.

7. Valence bond theory

Valence bond theory is a quantum mechanical approach to describe bonding in molecules. A covalent bond is a region of high electron charge density that results from the overlap of atomic orbitals between two atoms.

8.A. Hybridization of atomic orbitals

VB theory predicts bonding by overlap of atomic orbitals. There are two easily identified problems with this. First the bonding of C would predict only two bonds since there are only two half-filled orbitals. We know that carbon often forms four bonds. This part of the problem can be solved by "promoting" a 2s electron to an empty 2p orbital. The second problem is that from orbital shapes we would predict that the three p orbitals used to overlap in bonding would be at 90° and that the bonds formed from overlap of the p orbitals and the half-filled s orbital would be different. We know from experiment that the molecular geometry is tetrahedral and that all four bonds are equivalent. We therefore say that all four overlapping orbitals are equivalent and are the result of a mixing or hybridizing of the s and p orbitals. Because these new hybridized orbitals result from mixing one s and three p orbitals, they are designated sp^3. We end up with the same number of orbitals that we started with, namely four.

8.B. Hybrid orbitals involving d subshells

The maximum number of hybrid orbitals involving s and p orbitals is four, which will accommodate eight electrons. For molecules in which the central atom has an expanded octet, such as PCl_5, d subshell orbitals must be used. In this case, the orbitals will be one 3s, the three 3p orbitals and one 3d orbital. The hybridization is described as sp^3d. Do not confuse the superscript with the number of electrons in each orbital. In denoting the hybrid orbitals we are using the superscripts to describe the number of each type of orbital mixed for the hybrid orbital. In the case of sp^3d, there will be a total of five hybrid orbitals and we can accommodate ten electrons.

9. Predicting hybridization and orbital overlap in a molecule

In predicting the hybridization of a molecule it is useful to blend VSEPR and valence bond theory. The procedure is as follows:

A. Write a plausible Lewis structure

B. Use VSEPR method to predict the electron-pair geometry of the central atom.

C. Select the hybridization that corresponds to the predicted geometry.

D. Describe the orbital overlap.

Exercise (text) 10.6B. Describe a hybridization scheme for the central atom and the molecular geometry of the triiodide ion, I_3^-.

Solution: First we must write the Lewis structure. The skeleton structure is:

I--I--I

There are 3 x 7 + 1 = 22 electrons. Four of these are used in the single bonds. The other electrons are distributed around the I atoms to give completed octets. However, one pair of electrons is left. This can be assigned to the central atom as an expanded octet. This electron distribution will predict a trigonal bipyramid electron distribution (AX_2E_3). This geometry will require sp^3d hybridization. The molecular geometry is linear.

10. Hybridization used in forming multiple bonds

For multiple bonds, we can use the geometry to predict the hybridization. The multiple bond consists of a sigma bond, which is end-to-end overlap of orbitals, and pi bonds. A pi bond is the side-by-side overlap of p orbitals. A double bond will have one sigma and one pi bond and a triple bond will have one sigma and two pi bonds.

Example 10.2. Predict the hybridization and geometry of HCOOH.

Solution: The skeleton structure is:

```
        Oa
        |
    Ha--C--Ob--Hb
```

where the superscripts are used to distinguish the atoms. The central C atom should have trigonal planar geometry since the double bond is used like a single bond for the purposes of determining the geometry. The C atom would have sp^2 hybridization. O^a would overlap a p orbital with the sp^2 of C, and have one p orbital for the multiple bond. The O^b should have tetrahedral electron geometry. We would predict sp^3 hybridization and both H atoms to overlap with 1s orbitals. The bond angles around the C, H-C-O and O-C-O would be about 120°. The O^b would have a bond angle, C-O-H of less than 109° because of the two lone pairs.

11. Geometric isomers

Because of the side-by-side overlap of the p orbitals to form pi bonds in multiple bonds, rotation of the atoms about these bonds does not occur. This leads to geometric isomers. An example is $C_2H_2Br_2$

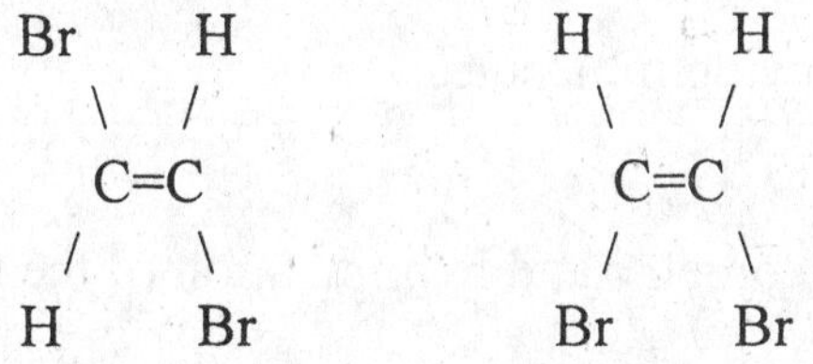

```
Br    H           H    H
  \  /             \  /
   C=C              C=C
  /  \             /  \
H     Br          Br   Br
trans             cis
```

Both of these isomers would be named 1,2 dibromoethene. To distinguish the two isomers, we add the prefix trans or cis. The trans isomer has Br on opposite sides of the double bond, that is, the Br are on opposite sides of a plane running through the two carbon atoms and perpendicular to the plane of the atoms in the molecule, while the cis prefix refers to both Br groups on the same side of the double bond. (The first isomer is trans, the second, cis.)

Example 10.3. **(a)** Predict a plausible molecular geometry for methanol, CH_3OH; **(b)** propose a hybridization scheme for the central atom that is consistent with the predicted geometry; and **(c)** sketch a bonding scheme for the molecule.

Solution:

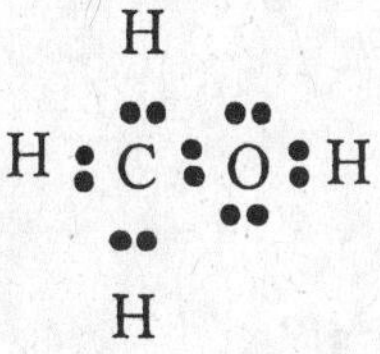

```
    H
    ••  ••
H : C : O : H
    ••  ••
    H
```

The central C atom is predicted to have tetrahedral geometry and therefore will have sp^3 hybridization. The O will also have sp^3 hybridization since the two unshared pairs of electrons will give the C-O-H bond bent geometry. The bond angle will be 109° because of the tetrahedral electron geometry. The bonding scheme will be an overlap of sp^3 orbitals of C with 1s orbitals of H. The C-O bond will be an overlap of two sp^3 orbitals, one from C and one from O. The O-H bond will be an overlap of the O sp^3 with the H 1s orbital.

12. Molecular orbitals

Molecular orbitals more accurately describe some of the properties of molecules than atomic orbitals. Acceptable solutions to wave equations written for the electrons in a molecule are called molecular orbitals (just as solutions to wave equations for electrons in an atom are called atomic orbitals). There are two types of MOs, bonding and anti-bonding. Electrons can be assigned to MOs by a process similar to the Aufbau process for filling atomic orbitals.

13. Benzene and aromatic compounds

Bonding in benzene can most easily be explained by a combination of valence bond theory for the sigma bonds and molecular orbital theory for the pi bonds.

Quiz A

1. Describe the molecular geometry of the following molecules:

   a. $BeCl_2$ b. $SnCl_2$ c. XeF_2

2. Describe the electron geometry, molecular geometry and hybridization of the central atom in ClF_3.
3. A mathematically based hypothesis that explains why all bond distances and angles in a methane molecule are equivalent is:
   a. delocalization of electrons
   b. resonance
   c. electronegativity
   d. hybridization
   e. none of these
4. Describe the orbital overlap to form a pi bond.
5. A multiple bond
   a. is weaker than a single bond
   b. is longer than a single bond
   c. can only be formed by s orbitals
   d. has both sigma and pi bonds.
6. Define bond order.
7. What is the maximum number of electrons in a molecular orbital?
8. Define a covalent bond according to valence bond (VB) theory.
9. The combination of s and p orbitals to form sp^3 hybrid orbitals results in how many hybrid orbitals?

Quiz B

1. Which of the following contains an sp^2-sp^2 sigma bond?
   a. C_2H_4
   b. C_2H_6
   c. HCN
   d. CH_3OH
2. Describe the molecular geometry of the following molecules:

   a. PCl_5 b. H_2O c. CO_2

3. Which of the following does not have a molecular dipole moment?

a. HCl b. CO c. BCl_3 d. H_2O e. none of these

4. Generally, short bonds are
   a. ionic bonds
   b. strong bonds
   c. between like atoms
   d. polar bonds
   e. none of these
5. Describe the hybridization of the C atom in CCl_4.
6. Describe a combination of atomic orbitals according to valence-bond theory that will result in five equivalent hybrid orbitals.
7. Are antibonding molecular orbitals higher or lower in energy than corresponding bonding molecular orbitals?
8. If two atomic orbitals are combined, how many molecular orbitals are formed?

Quiz C

True/False

1. The type of hybrid orbitals in a central atom can be determined by the geometry of the molecule.
2. The first row elements can expand their valence-shell beyond an octet.
3. Polar bonds are formed by atoms with different electronegativities.
4. A tetrahedral arrangement of electron clouds always implies sp^3 hybridization.
5. A σ bond is formed by the side-by-side overlap of p orbitals.
6. The number of molecular orbitals formed is equal to one-half the number of atomic orbitals combined.

Problems/short answer

7. Predict the molecular geometry of H_2F^+.
8. Predict the hybridization in a trigonal bipyramidal arrangement of electron clouds.
9. Describe the bonding in benzene. What is the hybridization of the carbon atoms?

Self Test

Multiple Choice

1. The geometric shape of $SnCl_4$ is
   a. linear
   b. bent
   c. square planar
   d. tetrahedral
   e. trigonal planar

2. In molecular orbital theory
   a. the number of orbitals is equal to the number of atomic orbitals used to generate them.
   b. each molecular orbital contains four electrons.
   c. a stable molecule contains more electrons in bonding orbitals than nonbonding orbitals.
   d. all of the above are true.
   e. none of the above are true.
3. Hybridization
   a. occurs by mixing orbitals on a single atom in a molecule.
   b. occurs between orbitals of very different energies.
   c. results in more orbitals than the orbitals mixed together.
   d. results in hybrid orbitals of different shapes.
4. The following molecule has the largest dipole moment:
   a. BCl_3
   b. CO_2
   c. CH_4
   d. CO
   e. NH_3

Short Answer

5. Ozone, O_3, is a nonlinear molecule with an O-O-O bond angle of 117^0. Explain this observation.
6. Use VSEPR theory to predict the shapes of the following molecules:
   a. SiH_4 b. Cl_2 c. $TeCl_2$
7. What types of hybrid bonds can be formed by elements in the third period that cannot be formed by elements in the second period?
8. Predict the stability of the He_2 molecule from MO theory. What is the bond order?
9. Predict the hybridization of the C atoms in CH_2CH_2.
10. Use MO theory to predict whether N_2 is stable and if it would be paramagnetic.

Fill in the blank

11. Molecular shape is determined by ______________________ around the atoms.
12. ____________ is a bond formed by head on overlap of orbitals and has its shared electrons centered about the axid between the two nuclei.
13. ________________ are the combination of wave functions for atomic orbitals which form a new set of equivalent wave functions.

14. Electrons in ______________________________ do not occupy the central region between the nuclei and do not contribute to bonding.
15. ____________________ = 1/2 (# of bonding electrons – # of antibonding electrons).

CHAPTER 11

Liquids, Solids, and Intermolecular Forces

With solids and liquids, intermolecular forces play an important role in their physical and chemical properties. This chapter will examine different types of forces in solids and liquids.

Chapter Objectives: You should be able to:

1. Describe intermolecular forces.
2. Define vaporization and the enthalpy of vaporization.
3. Define vapor pressure and know the relationship between vapor pressure and boiling point.
4. Define critical point, critical temperature and critical pressure.
5. Describe phase changes between a liquid and solid.
6. Define sublimation and relate the enthalpy of sublimation to the enthalpy of fusion and the enthalpy of vaporization.
7. Determine the state of a substance from a phase diagram and determine the triple point
8. Describe dispersion forces and the various types of dispersion forces.
9. Know which properties are affected by intermolecular forces
10. Define a network covalent solid and give an example
11. Relate the attractive forces in an ionic solid to the charge on the ions and the ionic radii.
12. Describe body-centered and face-centered crystal lattices.
13. Know why ionic crystal structures are more complicated than crystals of single atoms.
14. Describe the experimental determination of crystal structures.

Chapter Summary

1. Intermolecular forces

Intermolecular forces are forces between molecules. They determine the physical properties of liquids and solids. Intermolecular forces are likely to have the greatest effect at low temperatures and high pressures. Under these conditions, attractive forces are more likely to overcome the tendency of molecules to stay apart because the thermal energy is less and the molecules are closer together.

2. Vaporization and the enthalpy of vaporization

Vaporization is the conversion of liquid to a gas. Enthalpy of vaporization is the amount of heat that must be absorbed to vaporize a given quantity of liquid at a constant temperature.

Example 11.1. Condensation is the reverse of vaporization. How does the enthalpy of vaporization relate to the enthalpy of condensation? Write the appropriate equations.

Solution: The equation related to the enthalpy of vaporization is $H_2O(l)$ --> H_2O (g) and for condensation the reaction is reversed. Therefore the $\Delta H_{vap} = -\Delta H_{condensation}$.

Exercise (text) 11.1A. To vaporize a 1.50 g sample of liquid benzene, C_6H_6 requires 652 J of heat. What is ΔH_{vap} of benzene in kJ/mol?

Solution: The strategy is to convert 1.50 g to moles and to convert J to kJ. It requires 652 J to vaporize the number of moles contained in 1.50 g. We can convert this to the number of J required for one mole.

$$\frac{652 \text{ J}}{1.50 \text{ g}} \times \frac{78.1 \text{ g}}{1 \text{ mol}} \times \frac{1 \text{ kJ}}{1000 \text{ J}} = 33.9 \text{ kJ/mol}$$

3. Vapor pressure and the relationship between vapor pressure and boiling point

Vapor pressure of a liquid is the pressure exerted by the vapor in dynamic equilibrium with a liquid at room temperature. The rate of condensation and vaporization are equal when this equilibrium is reached. At the molecular level, changes are occurring, but overall the concentration of the vapor in the gas phase stays the same.

The boiling point of a liquid is the temperature at which the vapor pressure of the liquid is equal to the atmospheric pressure. The temperature of a liquid will not increase at the boiling point because the energy will go into converting more liquid to vapor, not raising the temperature of the liquid. For this reason, it is difficult to cook at high altitude. The temperature of the boiling water is below the normal boiling point (the boiling point at 1 atm of pressure).

Example 11.2. Suppose that the equilibrium illustrated in Figure 11.3 is between liquid water and its vapor at 22° C. What mass of water vapor would be present in a vapor volume of 275 mL at 22° C? Use data from Table 11.2.

Solution: From Table 11.2 we get the vapor pressure of water at 22°. This gives us the volume, temperature and pressure. Assuming the water vapor behaves as an ideal gas, solve for n, the number of moles and obtain the mass:

$$n = PV/RT = \frac{19.8 \text{ mmHg} \times \frac{1 \text{ atm}}{760 \text{ mmHg}} \times 0.275 \text{ L}}{0.08206 \text{ Latm mol}^{-1} \text{K}^{-1} \times 295 \text{K}} = 2.96 \times 10^{-4} \text{ mol}$$

Converting to mass gives:

$$2.96 \times 10^{-4} \text{ mol} \times (18.0 \text{ g/mol}) = 5.33 \times 10^{-3} \text{ g}$$

4. Critical point, critical temperature and critical pressure

The critical temperature is the temperature above which vapor cannot be liquefied no matter what pressure is applied. The critical pressure is the minimum pressure required to produce liquefaction at the critical temperature. The critical point is the point on a phase diagram that is the end point of the liquid-vapor line.

5. Phase changes between a liquid and solid

Melting or fusion is the process of a solid becoming liquid. Freezing is the reverse of melting, that is, a liquid becoming solid. The quantity of heat required to melt a given amount of solid is called the enthalpy of fusion. Melting is an endothermic process.

Example 11.3. Which of the substances in Table 11.4 in the text requires the greatest quantity of heat to melt a 1.0 g sample? Which requires the smallest quantity?

Solution: To answer this question, we need to look at the enthalpy of fusion. Since melting is an endothermic process, the substance with the largest heat of fusion, Fe, will require the most heat. The substance with the smallest heat of fusion, Hg, will require the least heat. The heat of fusion is the enthalpy for the change in going from solid to liquid. However, these heats are for a mole of substance and we are asked to estimate per gram. If we consider the molecular weights, Hg will have the smallest number of moles and since the enthalpy of fusion is also small, it will require the least heat. Water will have the most moles per gram based on its molecular weight. There will be approximately three times as many moles of water as iron. Since the enthalpy of fusion of water is more than a third that of iron, it will take more heat to melt 1.0 g of water than 1.0 g of iron.

6. Relating the enthalpy of sublimation to the enthalpy of fusion and the enthalpy of vaporization

Sublimation is the process of going directly from a solid to a gas. The enthalpy of sublimation is the sum of the enthalpy of fusion and the enthalpy of vaporization.

Example 11.4. Write equations to describe the sublimation of water and show the relationship between the enthalpy of sublimation, vaporization, and fusion.

Solution: Write the equations for the reactions that are occurring. The sublimation of water would be $H_2O\ (s) \dashrightarrow H_2O\ (g) \qquad \Delta H_{subl}$

This can be rewritten as the sum of two equations and ΔH calculated by Hess's Law:

$H_2O(s) \dashrightarrow H_2O(l) \quad \Delta H_{fusion}$

$H_2O(l) \dashrightarrow H_2O\ (g)\ \Delta H_{vap}$

Summing these two equations gives:

$H_2O\ (s) \dashrightarrow H_2O\ (g) \qquad \Delta H_{subl} = \Delta H_{fusion} + \Delta H_{vap}$

7. Triple point and phase diagrams.

The triple point is the only temperature and pressure at which solid, liquid and gas can coexist.

A phase diagram is a plot of temperature vs pressure. Lines are drawn to represent the equilibrium between phases. The boiling point of a liquid will increase with pressure. When this is plotted on a phase diagram, at pressures and temperatures which fall below the line, the substance will be a gas. It is useful to apply a common sense approach to interpreting phase diagrams -- at high temperature and low pressure substances will usually be in the gas phase (lower right on a P vs T diagram), while at low temperatures and high pressure substances will usually be in the solid phase (upper left of a P vs T diagram). Intermolecular forces are important in determining the physical properties of a substance, including whether it is solid, liquid, or gas under a given set of conditions.

8. Dispersion forces

Dispersion force is the attractive force between two molecules resulting from the instantaneous dipole of one molecule and the induced dipole of the other. Polarizability is a measure of the ease with which electron charge density is distorted by an external electric field. How does the polarizability of a molecule relate to size? In large molecules the outer electrons are more loosely bound and can shift toward another atom more easily than in small molecules.

A. Dipole-dipole forces

Remember that dipole-dipole forces result from a permanent dipole and are in addition to dispersion forces. For this reason, polar molecules have greater intermolecular forces for a given molar mass.

Exercise (text) 11.6A. Of the two substances BrCl and IBr, one is a solid at room temperature and the other is a gas. Which is which? Explain.

Solution: Dispersion forces become stronger with increasing molecular weight so we would predict IBr would have stronger dispersion forces and would be a solid.

B. Hydrogen bonds

A hydrogen bond is a strong dipolar intermolecular force between a polarized hydrogen atom in one molecule and a non-bonded "lone" pair on a nitrogen, oxygen, or fluorine atom in a second molecule. Hydrogen bonding is important for many life processes. These include protein and nucleic acid structure as well as the properties of water.

Exercise (text) 11.8. For each of the following substances, comment on whether hydrogen bonding is an important intermolecular force:

a. NH_3 **b.** CH_4 **c.** C_6H_5OH

d. CH_3COOH **e.** H_2S **f.** H_2O_2

Solution:

a. Yes, H is bonded to a highly electronegative nonmetal, N.

b. No. H is not bonded to N, O, or F.

c. Yes. H is bonded to oxygen, a highly electronegative nonmetal. The H on the benzene ring will not be involved in H bonding.

d. Yes. H is bonded to oxygen. The CH_3 hydrogens will not be involved in H bonding.

e. No. S is not N, O, or F.

f. Yes. H is bonded to oxygen.

9. Properties affected by intermolecular forces

Intermolecular forces affect properties such as boiling point, melting point, heat of vaporization, and surface tension. Surface tension is the amount of work required to extend a liquid surface.

10. Network covalent solid

A network covalent solid is one in which covalent bonds extend throughout the crystalline solid. Examples are graphite and diamond.

11. Relating the attractive forces in an ionic solid to the charge on the ions and the ionic radii

In an ionic solid, each ion is simultaneously attracted to several ions of the opposite charge. The attractive forces between a pair of oppositely charged ions increases as the charges on the ions increase and as the ionic radii decrease.

Exercise (text) 11.9. Arrange the following ionic solids in the expected order of increasing melting point: KCl, MgF_2, KI, and CsBr.

Solution: Mg^{2+} has a higher charge than the other cations. It is also smaller than Cs or K. Therefore, MgF_2 will have the highest melting point. K is smaller than Cs and Cl is smaller than Br. KCl will have a higher melting point than CsBr. This gives the order CsBr < KCl < MgF_2. Where does KI fit in? KI will have a lower melting point than KCl because I is larger. I is also larger than Br, however Cs is larger than K. Which has a lower melting point, CsBr or KI? The difference in size between Cs and K is greater than the difference in size between I and Br, so we would expect CsBr < KI <KCl < MgF_2.

12. Body centered and face centered crystal lattices

Two of the common crystal lattices are body-centered cubic and face-centered cubic. These can be described as a box with eight atoms or molecules at the corners (1/8th of each atom is within the box) and an additional atom in the center entirely within the box in the case of body-centered cubic or with additional atoms on the faces of the cube (1/2 within the box) in the case of face centered cubic.

Example 11.5. Show that the number of atoms in the fcc unit cell is 4.

Solution: The atoms at the corners of the cube are shared among eight unit cells. Only one-eighth "belongs" to each unit cell. The eight corners give a total of one atom. The atoms in the face of the cube are shared by two unit cells so that one-half of each atom belongs to the unit cell. There are six faces for a total of three atoms. The three atoms from the faces and the one atom from the corners of the unit cell give a total of four atoms.

Example 11.6. The edge length of the unit cell of an iridium crystal is 3.833 Å. Iridium crystallizes in a face-centered cubic unit cell. Calculate the atomic radius of the iridium atom.

Solution. For a face centered cubic cell, the length of the diagonal is four times the radius of the atom, and $(4r)^2 = l^2 + l^2$ where l is the length.

$$4r = \sqrt{2}\ l$$

$$r = \frac{\sqrt{2}\ (3.833\ \text{Å})}{4} = 1.355\ \text{Å} = 13.55\ \text{nm}$$

12. Ionic crystal structures more complicated than crystals of single atoms

Ionic crystal structures are more complicated because there are two types of ions present. The sizes of the cations are different from the sizes of the anions. A unit cell of an ionic crystal must generate the entire crystal by straight-line displacements in three dimensions, indicate the coordination numbers of the ions, and be consistent with the formula of the compound.

13. The experimental determination of crystal structures

Crystal structures are determined by measuring the scattering of x-rays. By measuring the angle, θ, at which scattered x-rays of known wavelength have their greatest intensity, we can calculate the spacing, d, between atomic planes from the formula

$$n\lambda = 2d \sin\theta$$

Skills Test Problem

For each of the following pairs, chose the lower boiling member. Explain.

a. H_2 or O_2

b. NH_3 or PH_3

c. PH_3 or AsH_3

d. SO_2 or SiO_2

Solution:

a. H_2 lower dispersion forces

b. PH_3, no H bonds

c. PH_3, lower dispersion forces

d. SO_2, molecular vs. network covalent bonds

Quiz A

1. Which of the following would you expect to have the highest boiling point: C_3H_8, CO_2, or CH_3CN?
2. Carbon tetrachloride, CCl_4, has a lower enthalpy of vaporization than water. Explain.
3. A vapor is in equilibrium with its liquid. Some of the vapor escapes. The result is:
   a. condensation rate increases
   b. condensation rate decreases
   c. vaporization rate increases
   d. vaporization rate decreases
4. The enthalpy of fusion is the enthalpy for the process:
   a. solid going to gas
   b. solid going to liquid
   c. liquid going to solid
   d. gas going to liquid
   e. liquid going to gas
5. How many atoms are in a body centered cubic unit cell?
6. Describe the bonding in a network covalent solid.
7. Which of the following will exhibit hydrogen bonding?
   a. NH_3
   b. CH_4
   c. O_3
   d. none of these
   e. all of these
8. Which of the following would you expect to have a higher boiling point: hexane, C_6H_{14}, or 1-hexanol, $C_6H_{13}OH$? Why?
9. Define sublimation.

10. ΔH_{vap} of methanol, CH_3OH, is 38.0 kJ/mol. How much heat is required to vaporize 100.0 g of methanol at 25 °C?

Quiz B

1. Which of the following will have the highest boiling point: NH_3, H_2O, or HF?
2. Which of the following will have the lowest melting point: NaF, NaBr, NaCl, or NaI?
3. When the vapor pressure of a liquid is equal to atmospheric pressure, the temperature of the liquid is:
   a. the sublimation point
   b. the critical point
   c. the normal boiling point
   d. the decomposition point
   c. the normal boiling point
   e. none of these
4. How many atoms are in a simple cubic unit cell?
5. Which of the following factors will determine the vapor pressure of a liquid?
   a. hydrogen bonding
   b. London forces
   c. polarity of the molecule
   d. molar mass
   e. all of these
6. For NaCl, ΔH_{vap} = 181.8 kJ/mol and ΔH_{subl} = 229.0 kJ/mol. What is ΔH_{fus} for NaCl?
7. Which of the following compounds would you expect to have the highest melting point: MgF_2, NaF, HCl, or CH_4?
8. Describe the condition known as supercooling.
9. Define polarizability.

Quiz C

True or False

1. Dispersion forces are some of the strongest intermolecular forces.
2. Dipole-dipole forces are generally weak and are significant only when the molecules are in close contact.
3. Small atoms with fewer electrons are highly polarizable.
4. Hydrogen bonding can occur between CH_4 and H_2O.
5. Hydrogen bonding gives rise to higher boiling points for small molecules than might be predicted from molecular mass and dipole character.

Multiple Choice

6. Graphite is an example of:
   a. an ionic compound
   b. a polarizable molecule
   c. a covalent network solid
   d. a hydrogen bond acceptor
7. Allotropes
   a. have the same physical and chemical properties
   b. are different structural forms of the same element
   c. are constituted of particles that are randomly arranged and have no ordered structure
   d. are polar molecules
8. From a phase diagram, one can determine:
   a. the crystal structure of a compound
   b. the boiling point
   c. the hydrogen bonding in a compound
   d. the molarity of the solution
9. The effect of pressure on the slope of the solid-liquid boundary line in a phase diagram depends on:
   a. the relative densities of the solid and liquid phases
   b. the sublimation point
   c. the triple point
   d. the melting point
10. The structure of crystals is often determined by:
   a. careful microscopic investigation
   b. phase diagrams
   c. x-ray diffraction
   d. surface tension

Self Test

Matching

1. allotrope	a. conversion of a gas to a liquid
2. boiling point	b. The direct passage of molecules from the solid state to the vapor state
3. polarizability	c. conversion of a solid to a liquid
4. hydrogen bond	d. the quantity of heat required to melt a given quantity of solid
5. phase diagram	e. a pressure-temperature plot indicating the conditions under which a substance exists as a solid phase, a

liquid, or a gas, or some combination of these in equilibrium

6. sublimation	f. one of two or more forms of an element that differ in their basic molecular structure.
7. fusion	g. a measure of the ease with which electron charge density in an atom or molecule is distorted by an external electric field.
8. condensation	h. the temperature at which the vapor pressure of the liquid is equal to atmospheric pressure.
9. enthalpy of fusion	i. The quantity of heat and work required to vaporize a given quantity of liquid at a constant temperature
10. enthalpy of vaporization	j. a type of intermolecular force in which a hydrogen atom covalently bonded in one molecule is simultaneously attracted to a nonmetal atom in a neighboring molecule.

Problems/Short Answer

11. The dipole moment of HCl is 1.03D and in HCN it is 2.99 D. Which compound would you predict to have the higher boiling point? Why?
12. Viscosity of a liquid is the resistance to flow. What conclusions can you make the viscosity as it relates to intermolecular forces?
13. Describe the types of intermolecular forces that exist in HBr and H_2O.
14. The amount of heat evolved when 1.0 g of steam condenses at 100 ^{0}C is –2.27 kJ. What is the heat of vaporization of water?
15. In a gold crystal, the first order reflection (n = 1) is observed at $\theta = 22.20^0$ for x-rays of 204 pm. What is the distance between planes of the crystal?
16. The molar heat of fusion and vaporization of liquid potassium are 2.4 and 82.5 kJ, respectively. What is the molar heat of sublimation?

CHAPTER 12

Solutions: Some Physical Properties

In this chapter some of the physical properties of solutions will be discussed. The fundamental questions about solution formation and which substances dissolve in one another will also be considered.

Chapter Objectives: You should be able to:

1. Know the definitions of solute and solvent.
2. Know the definition of molarity, and how to calculate percent by mass, percent by volume, and mass/volume percent.
3. Express the concentrations of dilute solutions in ppm, ppb and ppt.
4. Calculate the molality of solutions and determine the amount of solute from the molality.
5. Define mole fraction, mole percent, and know how to calculate both quantities.
6. Relate intermolecular forces to solution formation and enthalpy of solution.
7. Describe the properties of an ideal and a nonideal solution.
8. Describe what is meant by dynamic equilibrium.
9. Describe the relationship between temperature and water solubility of ionic compounds and temperature and solubility of gases in liquids.
10. State Henry's law and know how to apply it.
11. Define colligative properties.
12. Know the relationship between vapor pressure of a mixture and mole fraction.
13. Calculate freezing point depression and boiling point elevation.
14. Know the definition of osmosis and the formula for osmotic pressure.
15. Assess the concentrations of species in solution to treat colligative properties of electrolyte solutions.
16. Define a colloid.

Chapter Summary

1. Solvent and solute

The solvent is the solution component that is present in the greatest mole fraction. The other components are the solute. Solutions can involve gases, liquids and solids as either or both solute and solvent.

2. Molarity, percent by volume and percent by mass

For solutions where both the solute and solvent are liquids, percent by volume is often used because volumes are so easily measured for liquids.

Molarity, M, is defined as:

$$M = \frac{\text{amt of solute (moles)}}{\text{volume of solution (L)}}$$

$$\%\ \text{mass} = \frac{\text{mass substan ce}}{\text{total mass}} \times 100\%$$

$$\%\ \text{volume} = \frac{\text{volume subs tan ce}}{\text{total volume}} \times 100\%$$

$$\text{mass/volume}\ \% = \frac{\text{mass subs tan ce}}{100\ \text{mL solution}} \times 100\%$$

Example 12.1. What is the percent by mass of a solution made by dissolving 163 g glucose in 755 g water?

Solution: To calculate the percent by mass, we want the grams of solute/gram of solution x 100%. The mass percent is:

$$\frac{163\ \text{g}}{163\ \text{g} + 755\ \text{g}} \times 100\ \% = 17.8\ \%$$

Example 12.2. Which of the following solutions has the greatest mass percent of solute?

a. 15.5 g NaCl in 100.0 g solution

b. 15.5 g Na_2SO_4 in 100.0 mL water

c. 155 g Na_3PO_4 per kg water

Solution: The nature of the compound does not matter in calculating the mass percent. The NaCl solution (a) will have the highest mass percent because the total solution (solute + solvent) is 100 g while the other two examples contain 100 g of solvent for the same amount of solute.

Example 12.3. What volumes, in milliliters, of ethanol and water would you mix to obtain 200 mL of a 40.0 percent-by-volume solution of ethanol in water? What assumption is required in your calculation?

Solution: To calculate the volume of ethanol in mL, multiply 200 mL by 0.400. This will give a volume of 80.0 mL. The remaining 120 mL will be water. This result is obtained by rearranging the definition of % by volume: % vol = 100 x (mL of solute/mL of solution). The volume of solution is the sum of the volume of ethanol and the volume of water. The assumption made in determining the volume of water is that mixing the volume of ethanol and the volume of water gives a total volume that is the sum of the two individual volumes.

3. Expressing the concentrations of dilute solutions in ppm, ppb and ppt

Exercise (text) 12.3A. What is the concentration in (a) ppb and (b) ppt corresponding to a maximum allowable level in water of 0.1 μg/L of the pesticide chlordane?

Solution: To obtain this level in ppb, express the numerator and denominator in the same units and then write the denominator as one billion times the unit chosen. If we choose the unit μg:

$$\frac{0.1\,\mu g}{1\text{ L x}\dfrac{10^3\text{ mL}}{\text{L}}\text{ x}\dfrac{1\text{ g}}{\text{mL}}\text{ x}\dfrac{10^6\,\mu g}{\text{g}}} = \frac{0.1\,\mu g}{10^6\text{ x}10^3\,\mu g} = 0.10\text{ ppb}$$

To obtain ppt, we need the denominator in units of 10^{12}. Multiply both the numerator and denominator by 10^3. The concentration in ppb is then 0.10 x 10^3 ppt or 100 ppt.

4. Calculate the molality of solutions.

Molality is defined as the number of moles of solute/kg of solvent and is a convenient concentration unit for many calculations involving physical properties of solutions. The important thing to remember is that the unit in the denominator is kg of solvent, not solution.

Example 12.4. What is the molality of a solution prepared by dissolving 5.00 mL of C_2H_5OH (d = 0.789 g/mL) in 75.0 mL of C_6H_6 (d = 0.877 g/mL?

Solution: First, remember the definition of molality. We need to determine the amount of solute in moles and the mass of solvent in kg. We use the density to determine the mass in grams in both cases. For the solute, we use the molar mass to deternnine the amount in moles. For the solvent, we need only convert to kg. Converting the solute first:

$$5.00\text{ mL x }\frac{0.789\text{ g}}{\text{mL}}\text{ x}\frac{1\text{ mol}}{46.0\text{ g}} = 0.0858\text{ mol}$$

Converting the solvent:

$$75.0\text{ mL x }\frac{0.877\text{ g}}{\text{mL}}\text{ x}\frac{1\text{ kg}}{1000\text{ g}} = 0.0658\text{ kg}$$

The molality is then:

$$\frac{0.0858\text{ mol solute}}{0.0658\text{ kg solvent}} = 1.30\text{ molal}$$

5. Mole fraction and mole percent.

The mole fraction , χ_i is

$$\chi_i = \frac{\text{amt component i (mol)}}{\text{total amount solution components (mol)}}$$

The mole % is obtained by multiplying χ x 100%.

Exercise (text) 12.6B. A 2.90 m solution of methanol (CH_3OH) in water has a density of 0.984 g/mL. What are the **a.** mass percent, **b.** molarity, and **c.** mole percent of methanol in the solution?

Solution: From the definition of molality and the information given, we can calculate the number of moles of methanol. Using the molar mass, calculate the number of grams to give the mass percent. Using the total mass, calculate the volume of solution for the molarity and finally, calculate the moles of solution to calculate the mole percent.

a. Calculation of mass percent:

2.90 moles methanol/kg of solvent from the definition of molality.

2.90 moles x 32.04 g/mol = 92.916 g methanol. This is the mass of methanol present in the solution. 1 kg of water is present. The total mass of the solution is then 1000 g H_2O + 92.916 g methanol = 1092.916 g solution. The mass percent is

$$\frac{\text{mass methanol (g)}}{\text{mass solution (g)}} \text{ x } 100\% = \frac{92.916}{1092.916} = 8.50\text{ \% methanol by mass}$$

b. Molarity is moles of solute/L of solution. We have determined that there are 2.90 moles of methanol/1092.9 g of solution. Converting the g of solution to L is accomplished by using the density as a conversion factor:

$$1.092.9\text{ g x } \frac{1\text{ mL}}{0.984\text{ g}} \text{ x } \frac{1\text{ L}}{1000\text{ mL}} = 1.11\text{ L}$$

The molarity is therefore:

$$\frac{2.90\text{ mol}}{1.11\text{ L}} = 2.61\text{ M}$$

c. The mole percent of methanol in the solution is calculated by using the total number of moles present. We have already calculated the moles of methanol. We also know, from definition, that there is 1 kg of solvent, water, or 1000 g. Using the molar mass of water, we determine the number of moles of water:

$$1000\text{ g x } \frac{1\text{ mol}}{18.02\text{ g}} = 55.5\text{ moles of water}$$

To find the mole percent use the moles of methanol and the total moles present:

$$\frac{2.90\text{ mol}}{(2.90\text{ mol methanol} + 55.5\text{ mol water})} \text{ x } 100\% = 4.97\text{ mol \%}$$

6. Enthalpy of solution

Whether solution formation is an endothermic or exothermic process depends on intermolecular forces of attraction. The comparative strengths of intermolecular forces between solute molecules, solvent molecules, and solute-solvent molecules are important in determining whether a solution forms between a given solute and solvent. Three possibilities are

a. All intermolecular forces are of comparable strength. We expect a solution to form because molecules of solute and solvent will mix randomly.

b. Intermolecular forces between solute and solvent molecules are stronger than other forces. A solution will form and we expect heat to be evolved.

c. Intermolecular forces between solute and solvent molecules are weaker than other intermolecular forces. A solution may form, but the process will be endothermic.

7. Ideal and nonideal solutions.

Nonideal solutions are characterized by intermolecular forces between solvent and solute molecules that are strong compared to the forces between individual components.

Ideal solutions have the following properties:

a. They are homogeneous solutions.

b. The volume of the solution is the sum of the volumes of the components mixed to form the solution.

c. ΔH for formation of the solution is zero.

Example 12.5. Following are four organic compounds. Rank them in order of increasing solubility in water. That is, list the one which dissolves to the smallest extent first, and so on.

a. acetic acid, CH_3COOH **b.** heptane $CH_3(CH_2)_5CH_3$

c. 1-octanol, $CH_3(CH_2)_6CH_2OH$ **d.** pentanoic acid, $CH_3(CH_2)_3COOH$

Solution: The least soluble is heptane because there are no hydrogen bonds and the molecule does not ionize. The second least soluble is 1-octanol because it has a long hydrocarbon chain and only one OH group for possible H bond interactions with water. Third is pentanoic acid. This molecule is smaller, can form H bonds between the O and H of water and will ionize to form ion-dipole interactions. The most soluble is acetic acid. Acetic acid has the same interactions as pentanoic acid, but has a much smaller hydrocarbon chain and will, therefore, be more soluble. The order of solubility from the least soluble to the most soluble is:

heptane < 1-octanol < pentanoic acid < acetic acid.

8. Dynamic equilibrium

Dynamic equilibrium exists when the rate of dissolving of a substance is equal to the rate of crystallization. A saturated solution has undissolved solute in dynamic equilibrium with the solution.

9. Solubility as a function of temperature

Most ionic compounds are more soluble in water with increasing temperature, however, most gases are less soluble in liquids with increasing temperature.

10. Henry's Law

Henry's Law states that at constant temperature, the solubility of a gas is proportional to pressure of the gas in equilibrium with the aqueous solution. This can be explained by the fact that the solubility equilibrium is established when the number of gas molecules entering the solution is equal to the number leaving the solution. When the pressure increases, the number of gas molecules per unit volume above the solution increases and the rate at which molecules enter the solution will increase.

Example 12.6. At 25 ° C and 1 atm gas pressure, the solubility of CO_2 is 149 mg/100 g water. When air at 25 °C and 1 atm pressure is in equilibrium with water, what is the quantity of dissolved CO_2 in mg/100 g water? Air contains 0.035 mole % $CO_2(g)$.

Solution: To solve this problem, determine the pressure of CO_2 in air under the conditions given. Then use Henry's Law to determine the concentration. Calculate the pressure of CO_2 by using the relationship of partial pressures and mole fractions for gaseous mixtures:

$$\frac{P_{CO_2}}{P_{total}} = \frac{n_{CO_2}}{n_{total}}$$

Since the mole percent of CO_2 is 0.035 %, the mole fraction is 0.035/100 = 0.00035. The atmospheric pressure is 1 atm and the pressure of CO_2 is 0.00035 x 1atm = 0.00035 atm. We now have the pressure of CO_2 and can apply Henry's Law.

$$k = \frac{S_{CO_2}}{P_{gas}} = \frac{149\text{ mg }CO_2 / 100\text{ g }H_2O}{1\text{ atm}} = 149\text{ mg }CO_2/100\text{ g }H_2O\text{ atm}$$

Using this value for k, calculate the solubility of CO_2 at the new pressure, 0.00035 atm.

$S = k \times P$ = (149 mg/100 g atm) x 0.00035 atm = 5.2×10^{-4} mg/g H_2O and multiplying by the quantity of water, 100 g, gives the solubility as 0.052 mg CO_2/100 g H_2O.

11. Colligative properties

Colligative properties are properties that depend only on the concentration of solute particles present, but not the nature of the solute. These formulas apply to dilute solution and solutions where the solute is nonvolatile.

12. Relationship between vapor pressure of a mixture and mole fraction

Raoult's Law, $P_A = X_A P_A°$, gives this relationship.

Example 12.7. The vapor pressure of pure water at 20.0 °C is 17.5 mmHg. What is the vapor pressure of water above a solution that is 1.00 m sucrose ($C_{12}H_{22}O_{11}$)?

Solution: Start from the definition of molality. A 1.00 m solution has one mole sucrose/1.00 kg H_2O. Calculate the moles of water in one kg and then calculate the mole fraction.

$$1.00 \text{ kg} \times \frac{1000 \text{ g}}{\text{kg}} \times \frac{1 \text{ mol}}{18.02 \text{ g}} = 55.5 \text{ moles}$$

The mole fraction of water is then:

55.5 moles/56.5 moles = 0.982

Multiplying the mole fraction times the vapor pressure of pure water gives

0.982 x 17.5 mmHg = 17.2 mmHg

Example 12.8. Above which of the following solutions does the vapor have the greater mole fraction of benzene at 25 °C: a solution with equal masses of benzene (C_6H_6) and toluene ($C_6H_5CH_3$), or one having equal numbers of moles of benzene and toluene?

Solution: The vapor with the greater mole fraction of benzene will be from the solution with the higher mole fraction of benzene. Because toluene has a higher molar mass than benzene, a solution that contains equal masses will have less toluene and more benzene on a mole basis. This solution will have a lower mole fraction of toluene and a lower vapor pressure of toluene. Therefore, the solution with equal masses of toluene and benzene will have a vapor with a higher mole fraction of benzene.

13. Freezing point depression and boiling point elevation.

The formulas for freezing point depression and boiling point elevation, ΔT_f and ΔT_b, respectively, are

$\Delta T_f = -K_f \times m$

$\Delta T_b = K_b \times m$

where m is the molality of the solution.

Example 12.9. What mass of sucrose, $C_{12}H_{22}O_{11}$, should be added to 75.0 g H_2O (K_b = 0.512 °Cm^{-1}) to raise the boiling point to 100.35 °C?

Solution: Calculate the molality of a solution that would have this boiling point and then calculate the mass in grams of sucrose required. Using the formula for boiling point elevation:

$\Delta T_b = K_b \times m$

The normal boiling point for water is 100 °C so (100.35 °C - 100 °C) = 0.35 °C = ΔT_b and from Table 12.2, K_b = 0.512 °Cm^{-1}. Substituting into the equation:

0.35 °C = 0.512 °C m^{-1} x m and rearranging,

m = 0.35 °C/0.512 °C m^{-1} = 0.68 m

0.68 m = 0.68 moles sucrose/kg solvent

Solving for the number of moles needed for 75.0 g

(0.68 moles sucrose/1000 g solvent) x 75.0 g solvent = 0.051 moles

Using the molar mass of sucrose to obtain the grams of sucrose:

0.051 moles x 342 g/mole = 17 g sucrose required to elevate the boiling point of 75.0 g water.

Exercise (text) 12.15A. A 1.065 g sample of an unknown substance is dissolved in 30.00 g of benzene (K_f = 5.12 °C m^{-1}, T_f = 5.53 °C); the freezing point of the solution is 4.25 °C. The substance is found to be 50.69% C, 4.23% H, and 45.08% O by mass. What is the molecular formula of the substance?

Solution: Using the formula for freezing point depression, determine the molality of the solution and from the molality the number of moles of the unknown. Then, using the mass of the unknown, calculate its molar mass. From the molar mass and the mass percent of each element, determine the molecular formula.

Using the formula for freezing point depression:

$$\Delta T_f = -K_f \times m$$

$$4.25\ °C - 5.53\ °C = -1.28\ °C = -5.12\ °C\ m^{-1} \times m$$

$$m = \frac{-1.28\ °C}{-5.12\ °C\ m^{-1}} = 0.250\ m = 0.250 \text{ moles substance/kg solvent}$$

Solving to find the amount in moles of the above solution

$$\frac{0.250 \text{ mol}}{1000 \text{ g}} \times 30.00 \text{ g} = 0.00750 \text{ moles}$$

This amount in moles came from 1.065 g of sample, therefore the molar mass of the unknown is 1.065 g/0.00750 moles = 142 g/mole.

Using the percent composition, find the number of moles of each element in the compound:

$$\frac{0.5069 \text{ g C}}{\text{g compound}} \times \frac{142 \text{ g compound}}{\text{mol compound}} \times \frac{1 \text{ mol C}}{12.0 \text{ g}} = \frac{5.998 \text{ mol C}}{\text{mol compound}} \sim 6 \text{ mole C}$$

$$\frac{0.0423 \text{ g H}}{\text{g compound}} \times \frac{142 \text{ g compound}}{\text{mol compound}} \times \frac{1 \text{ mol H}}{1.01 \text{ g}} = \frac{5.95 \text{ mol H}}{\text{mol compound}} \sim 6 \text{ mole H}$$

$$\frac{0.4508 \text{ g O}}{\text{g compound}} \times \frac{142 \text{ g compound}}{\text{mol compound}} \times \frac{1 \text{ mol O}}{16.00 \text{ g}} = \frac{4.00 \text{ mol O}}{\text{mol compound}} = 4.00 \text{ mole O}$$

The formula of the unknown is $C_6H_6O_4$.

14. Osmotic pressure

Osmosis is the flow of solvent into a solution through a semipermeable membrane, one which prohibits solute flow. When there is a difference in molarity of solutions on opposite sides of the membrane, there is a tendency for solvent molecules to preferentially cross the membrane from the side of lower solute molarity towards the side of higher solute molarity. Osmosis continues until the two solutions are of equal molarity. The formula for osmotic pressure is

$$\Pi = MRT$$

A semipermeable membrane is necessary to allow solvent molecules to pass through, but not solute molecules. If both solvent and solute molecules could pass through the membrane, there would be no pressure difference across the membrane.

Example 12.10. An aqueous solution is prepared by dissolving 1.08 g of human serum albumin, a protein obtained from blood plasma, in 50.0 mL of water. The solution has an osmotic pressure of 5.85 mmHg at 298 K. What is the molar mass of the albumin?

Solution: We are given the mass in grams of albumin in solution so we can work backward from the osmotic pressure equation to find the amount in moles. Knowing the amount in moles will then allow us to calculate the molar mass. The osmotic pressure formula is:

$$\pi = (n/V)RT$$

Rearranging to find the amount in moles

$$n = \pi V/RT$$

$$n = \frac{5.85\,\text{mmHg} \times \dfrac{1\,\text{atm}}{760\,\text{mmHg}} \times 0.0500\ \text{L}}{0.08206\,\text{L}\cdot\text{atm}\cdot\text{mol}^{-1}\cdot\text{K}^{-1} \times 298\ \text{K}} = 1.57 \times 10^{-5}\ \text{moles}$$

Notice that the pressure is converted to atm to be consistent with the units in the gas constant.

A sample of 1.08 g of albumin was used to make the solution. The molar mass of albumin is 1.08 g/1.57 x 10^{-5} moles = 0.688 x 10^{5} g/mol = 6.88 x 10^{4} g/mol.

15. Solutions of electrolytes

To assess the effects of electrolytes on colligative properties, we need to know the number of ions and molecules present. Strong electrolytes will ionize completely, while weak electrolytes will have both ions and molecules present.

Example 12.11. Which of the following aqueous solutions has the lowest and which has the highest freezing point: 0.010 *m* $CO(NH_2)_2$, 0.0080 *m* HCl, 0.0050 *m* $MgCl_2$, 0.0030 *m* $Al_2(SO_4)_3$?

Solution: To determine the freezing point depression for solutions that ionize, consider the number of particles. The first sample, $CO(NH_2)_2$, will not ionize and will therefore have a particle concentration of 0.010 *m* ; the second, HCl, will have 2 x 0.0080 *m* = 0.016 *m* ; the third, 3 x 0.0050 m = 0.015 m and the last; 5 x 0.0030 m = 0.015 m . The lowest particle concentration will be $CO(NH_2)_2$ and it will have the highest freezing point (smallest depression). The HCl will have the lowest freezing point because it has the highest concentration of particles and hence, the greatest freezing point depression.

16. Colloids

A colloid is a suspension of particles in a medium. The particles and medium can be either solid, liquid or gas. Some examples of colloids are milk, fog, hair spray, butter, opals, and starch solutions.

Quiz A

1. In a dynamic equilibrium involving a saturated solution:
   a. the rate of crystallization is equal to the rate of dissolving
   b. more solid dissolves into solution
   c. more solid crystallizes
   d. none of the above
2. Give two examples of colligative properties.
3. How much KCl would be added to water to make 100.0 g of aqueous solution that was 18% KCl by mass?
4. Which of the following has the lowest freezing point?
   a. 1.0 M NaCl
   b. 1.0 M $MgCl_2$
   c. 1.5 M CH_3OH
   d. 1.7 M glucose
5. Calculate the molality of a solution prepared by dissolving 35.0 g glucose, $C_6H_{12}O_6$, in 150 g of water.
6. A solution has a 2.00:5.00 mole ratio of benzene (C_6H_6) to toluene (C_7H_8). The vapor pressures of toluene and benzene at 25 °C are 28.4 mmHg and 95.1 mmHg, respectively. What are the partial pressures of the two hydrocarbons above the solution?
7. How would you prepare 250.0 g of an aqueous solution that is 4.60% KCl by mass?
8. Predict whether a mixture of benzene, C_6H_6, and water, H_2O, is likely to be heterogeneous or homogeneous. Why?
9. Doubling the pressure of a gas at constant temperature
   a. decreases the solubility
   b. doubles the solubility
   c. does not change the solubility
   d. increases the solubility by a factor of R x T, where R is the gas constant and T is the absolute temperature

e. none of these

Quiz B

1. What is the molality of an aqueous solution that is 22.0% NaCl by mass?
2. Which of the following aqueous solutions should have the highest boiling point?
   a. 0.010 M HCl
   b. 0.01 M $CaCl_2$
   c. 0.015 M glucose
   d. 0.02 M sucrose
3. The vapor pressure of pure water at 20.0 °C is 17.5 mmHg. What is the vapor pressure of water above a solution that is 0.200 m glucose? (Glucose is nonvolatile.)
4. What is the freezing point of an aqueous solution that is 0.30 m in CH_3OH?
5. Define osmotic pressure.
6. Calculate the molality of a solution prepared by dissolving 1.40 g Na_2CO_3 in 95.0 mL of water.
7. Would you expect CH_3OH to be soluble in water? Why?
8. Define supersaturated solution.
9. As temperature increases, gases
   a. become more soluble
   b. become less soluble
   c. do not have their solubility affected
   d. do not change solubility in a predictable fashion

Quiz C

True/False

1. Osmotic pressure is the migration of solvent through a semipermeable membrane.
2. Colligative properties depend on the amount of dissolved solute, but not on the identity of the solute.
3. When solute-solvent interactions are strong, the solute is likely to be soluble.
4. Ionic solvents are likely to dissolve in nonpolar solvents.
5. According to Henry's Law, increase in the pressure of a gas leads to a decrease in solubility.
6. When a solid is dissolved in liquid, the liquid is the solute.

Problems/Short Answer

7. A NaOH solution is 12.5% by weight. Calculate the molality.
8. Identify the intermolecular forces and predict whether a solution will form between CH_3CH_2OH and H_2O.

9. Calculate the osmotic pressure of a solution at 20 ^{0}C containing 12.0 g of glucose ($C_6H_{12}O_6$) in 1.00 L of solution.

Self Test

Match the following:

1. colligative property
2. solubility
3. molality
4. Raoult's Law
5. parts per million
6. Henry's Law
7. mole fraction
8. osmosis

a. states that the addition of a solute lowers the vapor pressure of the solvent and that the fractional lowering of the vapor pressure is equal to the mole fraction of the solute.

b. states that the solubility of a gas is directly proportional to the pressure maintained in the gas in equilibrium with the solution

c. the fraction of all the molecules in a homogeneous mixture contributed by that component

d. the amount of solute, in moles, per kilogram of solvent

e. the net flow of a solvent through a semipermeable membrane, from pure solvent into a solution or from a solution of a lower concentration into one of a higher concentration

f. refers to the concentration of the solute in a saturated solution

g. expresses the composition of a mixture as the number of parts of one component per million parts of the mixture as a whole

h. a physical property of a solution that depends on the concentration of solute in the solution but not the identity of the solute

Problems/Short Answer

9. Calculate the molarity and molality of concentrated HCl (36.5 percent by mass, density 1.18g/mL).
10. What is the boiling point of a solution with 2.3 g NaCl in 1.00 L of H_2O? (K_b = 0.52 ^{0}C/m)
11. Calculate the percent of NaCl by mass in a 1.3 m solution.
12. A solution contains 15.0 g of a nonelectrolyte dissolved in 200.0 mL of water. The freezing point of the solution is –0.150 ^{0}C.(K_f = 1.86 ^{0}Cm^{-1}) What is the molar mass of the compound?
13. Calculate the osmotic pressure of a solution at 25 ^{0}C containing 2.5 g of $MgCl_2$ in 200.0 mL of water.
14. The vapor pressure of water at 35 ^{0}C is 42.175 mm Hg. Calculate the vapor pressure of a solution that contains 1.2 g of NaOH in 100.0 mL of water.

Fill in the Blank

15. ___________________ contain particles with diameters in the range 2-1000 mm.
16. Three types of interactions among particles to take into account in determining the energy changes in solutions are ____________________, ________________, and ________________.
17. Gases become __________ soluble as the temperature is increased.
18. An increase in pressure of a gas leads to _______________ in solubility.

CHAPTER 13

Chemical Kinetics

Chemical reactions occur at very different rates. Often the rate of a reaction can be influenced by the conditions of the reaction. This chapter will consider the effects of a number of variables: concentrations of reactants, temperature, surface area and catalysis. In addition, in this chapter we will determine the concentration of reactants at various times of the reaction by using integrated rate equation. Experimental data can aid in determining whether a proposed mechanism is consistent with the observed data. We cannot prove a mechanism for a reaction, but only determine if the mechanism is a possible one for the observed reaction.

Chapter Objectives: You should be able to:

1. Name some of the variables that affect the speed of reactions.
2. Define rate of reaction.
3. Determine the instantaneous rate of reaction.
4. Define and determine the initial rate of reaction.
5. Write the rate laws of reactions and know the overall order of the reaction.
6. Determine the order of reaction and value of the rate constant by the method of initial rates.
7. Express the concentration of reactant as a function of time for a first order reaction: A--> Products.
8. Use quantities other than concentration to calculate rate constants and rate data.
9. Define half-life for a reaction.
10. Determine the rate of reaction and the integrated rate expression for zero-order and second-order reactions.
11. Describe a pseudo-first-order reaction and why it is often convenient to arrange reaction conditions to be pseudo-first-order.
12. Relate the rate of chemical reactions to the frequency of molecular collisions and effectiveness of the collisions.
13. Define activation energy.
14. Know the definitions of transition state, activated complex, and activation energy.
15. Know how to use the Arrhenius equation in calculations involving temperature, rate constants, and activation energy.
16. Know the important characteristics of an elementary process in a reaction mechanism.
17. Define and characterize a catalyst.

Chapter Summary

1. Variables that affect the speed of reactions

Some of the variables that affect the speed of reactions are concentration of reactants, temperature, and surface area. Reaction rates generally increase with the concentration of reactants and as the temperature increases. We will see later in this chapter how to quantitate these effects. For reactions that occur on a surface, the rate of reaction increases with increasing surface area.

2. Rate of reaction

The rate of a reaction refers to how much something changes in a unit of time. For chemical reactions, we will look at the rate of decrease of a reactant concentration and the rate of increase of a product concentration.

We can express the rate of change of concentration of a reactant, Δ[reactant]/Δ time, or of the product, Δ[product]/Δ time, but often we want to be more specific and relate the rate of change of concentration of one reactant to another or to a product. To do this, we use the stoichiometric coefficients for the balanced equation. If we use ν as the stoichiometric coefficient of the substance, the rate is defined as

$$\text{rate of reaction} = -\frac{\Delta[\text{reactant}]}{\nu_r \Delta t} = +\frac{\Delta[\text{product}]}{\nu_p \Delta t}$$

where ν_r and ν_p are the stoichiometric coefficients of reactant and product, respectively. This definition refers to the average rate of reaction. (See section 13.3.B for a definition of the initial rate of reaction.)

Example 13.1. For the reaction $CH_4 + 2\ H_2O \dashrightarrow CO_2 + 4\ H_2$, write the rate of reaction expression in terms of each reactant and product.

Solution: $\nu_{CH_4} = 1$, $\nu_{H_2O} = 2$, $\nu_{CO_2} = 1$ and $\nu_{H_2} = 4$. Therefore, the rate of reaction can be written

$$\text{rate of reaction} = \frac{-\Delta[CH_4]}{1 \text{ x } \Delta t} = \frac{-\Delta[H_2O]}{2 \text{ x } \Delta t} = \frac{+\Delta[CO_2]}{1 \text{ x } \Delta t} = \frac{+\Delta[H_2]}{4 \text{ x } \Delta t}$$

Example 13.2. For the reaction: $2\ NO_2(g) \dashrightarrow 2\ NO(g) + O_2(g)$, how does the rate of formation of [NO] relate to the rate of decrease of $[NO_2]$? How does the rate of formation of $[O_2]$ relate to the rates of change for $[NO_2]$ and [NO]?

Solution: Since the reaction stoichiometry indicates for each 2 moles of NO_2 reacted, 2 moles of NO form, the rate of decrease of $[NO_2]$ will be equal to the rate of increase of [NO]. Using the stoichiometric coefficients as in Example 13.1:

$$\frac{-\Delta[NO_2]}{2 \text{ x } \Delta t} = \frac{+\Delta[NO]}{2 \text{ x } \Delta t},$$

or canceling the 1/2,

$$\frac{-\Delta[NO_2]}{\Delta t} = \frac{+\Delta[NO]}{\Delta t}.$$

Note the difference in sign for the change in concentration with time of the reactant and product, since the amount of reactant is decreasing and the amount of product is increasing.

Similarly, 1 mole of $[O_2]$ will be produced for each 2 moles of $[NO_2]$. The rate of production of $[O_2]$ will be 1/2 that of the decrease in $[NO_2]$ or the production of [NO]. Therefore:

$$\frac{-\Delta[NO_2]}{2 \text{ x } \Delta t} = \frac{+\Delta[NO]}{2 \text{ x } \Delta t} = \frac{+\Delta[O_2]}{\Delta t}.$$

Exercise (text) 13.1B. In the reaction: 2 A + B --> 3 C + D, $-\Delta[A]/\Delta t$ is found to be 2.10 x 10^{-5} M s^{-1}.

a. What is the average rate of reaction of B?

b. What is the average rate of formation of C?

Solution:

a. Use the stoichiometry of the reaction to determine the rate of decrease of B. From the reaction stoichiometry, 2 moles of A react for each 1 mole of B. Therefore:

$$\frac{-\Delta[A]}{2 \text{ x } \Delta t} = \frac{-\Delta[B]}{\Delta t} = \frac{1}{2}(2.10 \text{ x } 10^{-5} \text{ M s}^{-1}) = 1.05 \text{ x } 10^{-5} \text{ M s}^{-1}$$

Another way to consider this is that for a given time period, say 1 sec, 2.10 x 10^{-5} moles of A will be used. Only 1/2 that amount of B will be used so the rate will be 1/2 that of A.

b. The rate of formation of C will be 3/2 the rate of change of A or 3.15 x 10^{-5} M s^{-1}. Note that the rate obtained for A is the rate of disappearance, and that for C, is the rate of formation.

3. Instantaneous rate of reaction.

The instantaneous rate of reaction is the rate of reaction at some particular time in the course of the reaction. It is derived from a tangent line to the concentration-time graph.

Exercise (text) 13.2A.

a. From Figure 13.5, determine the instantaneous rate of reaction at t = 300 s,

b. Use the result of (a) to calculate a value of $[H_2O_2]$ at t = 310 s.

Solution:

a. The instantaneous rate = -(slope of tangent line) = -(0 M - 0.63 M)/(570 s- 0 s) = 1.1 x 10^{-3} M s^{-1}. Note: it is generally more accurate to select points a large distance apart to determine the slope of a line. You must select points on the tangent line, not the concentration vs time curve itself.

b. The value of $[H_2O_2]$ is determined from the rate. If we take the time point as 310 s, and the $[H_2O_2]$ at that time as X, then:

Instantaneous rate = -(X - 0.63 M)/(310 s - 0 s) = 1.1 x 10^{-3} M s^{-1}

Solving for X

$X = -(1.1 \times 10^{-3}\ M\ s^{-1}) \times 310\ s + 0.63\ M = 0.29\ M$

We need to keep the signs straight here so that the concentration decreases. If we forget the minus sign, we would end up with more reactant after 310 s than we started with!

4. Initial rate of reaction

The initial rate of reaction is the rate of reaction at the start of the reaction and can be obtained by a line tangent to the first part of the concentration versus time curve (t = 0). Early in the reaction, the average and instantaneous rates of reaction are nearly equal.

Example 13.3. From the data below, determine **(a)** the initial rate of reaction and **(b)** the instantaneous rate of reaction at 100 s.

Reaction: A --> B

Time, s	[A], M
0	1.00
25	0.78
50	0.61
75	0.47
100	0.37
150	0.22
200	0.14

Solution:

a. We can determine the initial rate by using the first two data points.

$$\text{The initial rate} = \text{-slope} = -\frac{(1.00\ M - 0.78\ M)}{(0.0\ s - 25\ s)} = \frac{0.22\ M}{25\ s} = 8.8 \times 10^{-3}\ M\ s^{-1}$$

b. The instantaneous rate at 100 s can be calculated from the graph on which is plotted [A] vs time and taking the tangent, or we can estimate the rate by taking points on either side of 100 s over a narrow time interval and calculating the slope. Using this method and points at 75 s and 150 s the instantaneous rate is:

$$\text{Instantaneous rate} = \text{-slope} = -\frac{(0.22\ M - 0.47\ M)}{(150\ s - 75\ s)} = \frac{0.25\ M}{75\ s} = 3.3 \times 10^{-3}\ M\ s^{-1}$$

(Graphically, the slope yields a rate of $3.6 \times 10^{-3}\ M\ s^{-1}$.)

Notice that the instantaneous rate sometime after the start of the reaction is slower than the initial rate because the concentration of the reactant has decreased.

5. Rate laws of reactions

Rates of reaction can be written as

rate of reaction = $k[A]^m[B]^n$ ....

The proportionality constant, k, relates the rate of reaction to the concentration of reactants and is called the rate constant. The values of m and n must be determined by experiment, are generally small whole numbers, and are not necessarily related to the stoichiometric coefficients. The overall order of the reaction is the sum of the exponents (m + n). The order with respect to each reactant is the value of the exponent for that reactant.

Example 13.4. For the reaction 2 A + B --> C, the rate equation is:

rate of reaction = k[A][B]

What is the overall order of the reaction and the order with respect to A and B?

Solution: The overall order of the reaction is the sum of the exponents in the rate law. In this case, both concentrations are raised to the first power. The order of the reaction is 1 + 1 = 2. The order with respect to both A and B is first-order. The reaction is second-order overall.

6. Determining the order of reaction and value of the rate constant by the method of initial rates

One common method of determining the order of a reaction is by the method of initial rates. Remember that the initial rate is the instantaneous rate at the beginning of a reaction. By varying the starting concentrations of reactants and measuring the corresponding initial rates, we can determine the order of the reaction.

Suppose we have a reaction where rate of reaction = $k[A]^a[B]^b$. If we measure the initial rate of reaction at different starting concentrations of A and the same [B], then

$$\frac{\text{rate 1}}{\text{rate 2}} = \frac{k[A_1]^a[B_1]^b}{k[A_2]^a[B_2]^b}$$ and since $[B_1] = [B_2]$ and k is a constant,

$$\frac{\text{rate 1}}{\text{rate 2}} = \frac{[A_1]^a}{[A_2]^a} = \left\{\frac{[A_1]}{[A_2]}\right\}^a$$ where the exponent a is the only unknown.

Sometimes we can solve for a "in our heads," but a mathematical procedure is to take the log of both sides:

$$\log\frac{\text{rate 1}}{\text{rate 2}} = \log\frac{[A_1]^a}{[A_2]^a} = a\log\frac{[A_1]}{[A_2]}$$

Example 13.5. Initial rate data and corresponding concentration data are given below for the reaction: $NH_4^+(aq) + NO_2^-(aq) \text{-->} N_2(g) + 2\ H_2O(l)$.

Experiment	Initial $[NH_4^+]$	Initial $[NO_2^-]$	Rate, M s^{-1}
1	0.050 M	0.0050 M	0.69×10^{-7}
2	0.100 M	0.010 M	2.80×10^{-7}
3	0.100 M	0.0050 M	1.4×10^{-7}

Calculate the rate law and rate constant for this reaction.

Solution: We first write the general rate law:

rate of reaction = $k[NH_4^+]^a[NO_2^-]^b$

Then, using the method of initial rates, it is easiest to look for two experiments where one reactant has the same concentration and the other changes. Using experiments 1 and 3 we can solve for a.

$$\frac{\text{rate 1}}{\text{rate 3}} = \frac{k[0.050\ \text{M}]^a[0.0050\ \text{M}]^b}{k[0.100\ \text{M}]^a[0.0050\ \text{M}]^b}$$

Canceling terms and substituting in the rates gives:

$$\frac{\text{rate 1}}{\text{rate 3}} = \frac{[0.050\ \text{M}]^a}{[0.100\ \text{M}]^a}$$

$$\frac{0.69 \times 10^{-7}\ \text{M s}^{-1}}{1.4 \times 10^{-7}\ \text{M s}^{-1}} = \left(\frac{0.050\,\text{M}}{0.100\ \text{M}}\right)^a$$

$$0.50 = (0.50)^a$$

Therefore a = 1 and we know that

rate of reaction = $k[NH_4^+][NO_2^-]^b$

Using experiments 2 and 3 we now solve for b.

$$\frac{\text{rate 2}}{\text{rate 3}} = \frac{[0.010\ \text{M}]^b}{[0.0050\ \text{M}]^b}$$ since the other terms will cancel.

Substituting in the values for the rates gives:

$$\frac{2.80 \times 10^{-7}\ \text{Ms}^{-1}}{1.4 \times 10^{-7}\ \text{Ms}^{-1}} = \left[\frac{0.010\text{M}}{0.005}\right]^b$$

$2.00 = (2.0)^b$ and b = 1. The rate law is

rate of reaction = $k[NH_4^+][NO_2^-]$.

To solve for k, we can substitute concentrations and rates for any one of the experiments. Using experiment 1,

$$0.69 \times 10^{-7}\ \text{M s}^{-1} = k(0.050\ \text{M})(0.0050\ \text{M})$$

$$k = \frac{0.69 \times 10^{-7}\ \text{Ms}^{-1}}{(0.050\ \text{M})(0.0050\ \text{M})} = 2.8 \times 10^{-4}\ \text{M}^{-1}\ \text{s}^{-1}$$

7. Concentration as a function of time for a first order reaction.

For a first order reaction, the integrated rate equation is.

$$\ln \frac{[A]_t}{[A]_0} = -kt$$

Rearranging:

$$\ln [A]_t = -kt + \ln[A]_o$$

Notice that, for a first-order reaction, a plot of $\ln[A]_t$ vs time will give a straight line with a slope of -k.

Example 13.6. If the initial $[H_2O_2]$ = 0.882 M, calculate the time in the reaction at which $[H_2O_2]$ = 0.500 M. Use $k = 3.66 \times 10^{-3}\ s^{-1}$ for the first order decomposition of H_2O_2.

Solution: Use the integrated rate equation and solve for time. The initial concentration, $[A]_o$, is 0.882 M and the concentration at time t is 0.500 M. Using the above equation and solving for t gives:

$$t = \frac{(\ln[A]_o - \ln[A]_t)}{k}$$

Substituting in:

$$t = \frac{\ln(0.882) - \ln(0.500)}{3.66 \times 10^{-3}\ s^{-1}} = 155\ \text{sec}$$

8. Quantities other than concentration for calculation of rate constants and rate data

Rate constants and rate data can be calculated from data other than concentrations so long as the quantity is proportional to concentration. The most commonly used example is to use the pressure for gases in a reaction.

Example 13.7. A 45.0 g sample of N_2O_5 is allowed to decompose at 67 °C. What mass of N_2O_5 remains undecomposed after 5.00 min? Use the rate constant established in Example 13.5 (text).

Solution: In this case, we can show the mass is related to the number of moles and so long as the volume remains constant, the mass is related to the molarity. To solve the above equation,

$$\ln\frac{[N_2O_5]_t}{[N_2O_5]_o} = \ln\left[\frac{(\text{mass}_{[N_2O_5]_t})/(\text{Molar mass x volume})}{(\text{mass}_{[N_2O_5]_0})/(\text{Molar mass x volume})}\right]$$

$$= \ln\frac{(\text{mass}_{[N_2O_5]_t})}{(\text{mass}_{[N_2O_5]_0})} = -kt = -(5.83 \times 10^{-3}\ s^{-1}) \times 5.00\ \text{min} \times \frac{60\ s}{1\ \text{min}} = -1.749$$

$$\frac{(\text{mass}_{[N_2O_5]_t})}{(\text{mass}_{[N_2O_5]_0})} = 0.174$$

$mass_{[N_2O_5]_t} = 0.174 \times 45.0\ g = 7.83\ g$

9. Reaction half-life

For a first-order reaction, the half-life is $t_{1/2} = 0.693/k$.

Example 13.8. Use the result of Example 13.5 to determine the time required to reduce the quantity of N_2O_5 to 1/32 of its initial value.

Solution: We can solve this by using the integrated rate equation and setting $[A]_t = 1/32[A]_0$, but it is easier to recognize that 1/32 is an integral number of half-lives. In the first half-life, 119 s, the concentration is reduced to 1/2. In the second half-life the concentration is 1/4 the original concentration. Proceeding in this manner, in five half-lives, the concentration will be $\left(\frac{1}{2}\right)^5$ or 1/32. Five half-lives will be 5 x 119 s = 595 s.

10. Rate of reaction and the integrated rate expression for zero-order and second-order reactions

The rate of a zero order reaction is independent of concentration of reactants. For a zero-order reaction: A --> products, a plot of $[A]_t$ versus time will yield a straight line with a slope of -k and the rate of reaction will be constant over time (rate = k).

A second-order reaction has a rate equation in which the sum of the exponents is equal to two. For a second order reaction: A ---> products, the rate equation is

$$\text{rate of reaction} = k[A]^2$$

and a plot of $1/[A]_t$ versus time will yield a straight line with a slope of k. The integrated rate equation is

$$\frac{1}{[A]_t} = kt + \frac{1}{[A]_0}$$

Example 13.9. At what time after the start of the reaction, A --> products, will 1/3 of the intial starting material have decomposed, assuming the reaction is second-order, $k = 1.00 \times 10^{-2}\ M^{-1}\ s^{-1}$ and $[A]_0 = 2.00\ M$.

Solution: The integrated rate equation is

$$\frac{1}{[A]_t} = kt + \frac{1}{[A]_0}$$

If 1/3 of the material decomposed, 2/3 is left or $[A]_t = 2/3[A]_0$. Substituting into the integrated rate equation:

$$\frac{1}{\frac{2}{3}[A]_0} = kt + \frac{1}{[A]_0}$$

Rearranging:

$$\frac{1}{\frac{2}{3}[A]_0} - \frac{1}{[A]_0} = kt = \frac{1}{2[A]_0}$$

$$\frac{1}{2 \times 2.00\ M} = 1.00 \times 10^{-2}\ M^{-1}\ s^{-1} \times t$$

$$t = \frac{1}{2 \times 2.00\ M \times (1.00 \times 10^{-2}\ M^{-1}\ s^{-1})} = 25.0\ s$$

Example 13.10. Show that the next half-life period for a second order reaction is twice as long as the first half-life period,

Solution: For a second order reaction, the half-life is given by

$$t_{1/2} = \frac{1}{k[A]_0}$$

After the first half-life of the reaction, $[A]_t = 1/2[A]_0$. To calculate the second half-life, put in the new [A] which is 1/2 the first $[A]_0$. The second half-life will be:

$$t_{1/2} = \frac{1}{k[A]_t} = \frac{1}{k\frac{1}{2}[A]_0} = \frac{2}{k[A]_0}$$ or twice the first half-life.

11. Pseudo-first order reactions

Often kinetic treatment of reactions can be simplified by having one reagent in large excess. This can be used to analyze a second order reaction as a pseudo-first-order reaction. This treatment makes a reaction resemble a lower order reaction. This occurs because one reagent is in excess and the change in its concentration is very small over the course of the reaction.

12. Relating the rate of chemical reactions to the frequency of molecular collisions and effectiveness of the collisions

The rate of a chemical reaction is proportional to the product of the frequency of the molecular collisions and the fraction of the collisions that are effective ones. Kinetic-molecular theory provides a basis for calculating the fraction of molecules in a collection that possesses a certain kinetic energy. The fraction of molecules with enough energy to interact will increase with temperature.

13. Activation energy

The activation energy is the minimum total kinetic energy that two colliding molecules must bring to their collision for a reaction to occur.

14. Transition state theory

Collision theory does not explain many reactions. Transition state theory explains the rate of reaction by considering the activated complex that forms as the reaction occurs and reactants are transformed into products. Transition state theory considers the energy released when bonds are formed, and energy required to break bonds. The orientation of reactants required to form the activation complex is also considered.

15. Arrhenius equation

The Arrhenius equation is:

$$\ln k = -\frac{E_a}{RT} + \ln A$$

A more common way to use the Arrhenius equation is to measure the reaction rate at two different temperatures. The equation then becomes:

$$\ln\frac{k_2}{k_1} = \frac{E_a}{R}\left(\frac{1}{T_1} - \frac{1}{T_2}\right)$$

Example 13.11. For the reaction: $2\ N_2O_5(g) \dashrightarrow 4\ NO_2(g) + O_2(g)$, the rate constant at 332 K was $2.5 \times 10^{-3}\ M^{-1}s^{-1}$ and the E_a was 1.0×10^{-5} J mol^{-1}. Determine the temperature at which the rate constant, $k = 1.0 \times 10^{5}\ M^{-1}s^{-1}$.

Solution: From the above example, we know the rate of reaction at another temperature and the activation energy. We are asked to determine a second temperature given a second rate constant. Find the temperature by using the relationship between rate constants, activation energy, and temperature:

$$\ln\frac{k_2}{k_1} = \left(\frac{E_a}{R}\right)\left(\frac{1}{T_1} - \frac{1}{T_2}\right)$$

Substituting into the equation gives:

Solving for T_2 gives

$$\ln\frac{1.0\times10^{-5}}{2.5\times10^{-3}} = \left(\frac{1.0\times10^{5}\ \text{J/mol}}{8.3145\ \text{J mol}^{-1}\ \text{K}^{-1}}\right)\left(\frac{1}{332\ \text{K}} - \frac{1}{T_2}\right)$$

$$\ln(4.0 \times 10^{-3}) = 1.2 \times 10^{4} \times (1/332) - 1.2 \times 10^{4}(1/T_2)$$

$$-41.7 = -1.2 \times 10^{4}(1/T_2)$$

$$T_2 = 288\ \text{K}$$

16. Reaction Mechanisms

A reaction mechanism describes a chemical reaction as a series of elementary steps.
Keep in mind that these characteristics apply to elementary steps and not necessarily to reaction equations. The characteristics are:

1. The exponents of the concentration terms in the rate equation are the same as the coefficients of the balanced equation. For example, for the elementary reaction

   2 A + 3 B --> D + E

   Rate of reaction = $k[A]^2[B]^3$.

2. Unimolecular steps (single molecule dissociates) and bimolecular steps (two molecules collide) are much more likely than reactions involving simultaneous collisions of three or more molecules.

3. Elementary steps are reversible.

4. Species formed in one step and consumed in another are intermediates. They do not appear in the overall rate law or the net equation.

5. If one elementary step is much slower than others, it is the rate-limiting step.

Example 13.12. Consider the reaction between nitrogen monoxide and chlorine gases to form NOCl(g). The experimentally determined rate law is:

rate of reaction = $k[NO][Cl_2]$

A proposed mechanism is

Step 1: NO(g) + Cl_2(g) --> $NOCl_2$(g)

Step 2: $NOCl_2$(g) + NO(g) --> 2 NOCl(g)

a. Write the equation for the net reaction.
b. Which species is an intermediate?
c. Which step is likely to be the rate-determining step?

Solution:

a. The net equation is the sum of the equations for the two elementary steps:

2 NO(g) + Cl_2(g) --> 2 NOCl(g)

b. The intermediate is $NOCl_2$ because it does not appear in the net equation.

c. The rate-limiting step is the first step because the rate equation involves the concentration of both reactants. The rate equation does not involve the concentration of the intermediate. The rate equation for the elementary step is the same as for the overall reaction.

17. Catalysis

A catalyst increases the rate of a reaction without itself being changed by the reaction. The reaction rate is increased because of a lower activation energy. Homogeneous catalysis involves a reaction where substances are present within a homogeneous mixture. Heterogeneous catalysis occurs when all substances are not in the same mixture. This often happens when catalysts are on solid surfaces and the reactant molecules are absorbed. The product molecules are desorbed and freed from the solid surface.

Common catalysts in biological processes are enzymes. In enzyme catalysis, the reactants are called substrates. These substrates bind to the enzyme in a particular area called the active site. The enzyme-substrate complex dissociates into product molecules. Many drugs are inhibitors of enzymes; that is, they block the enzyme action, often by binding to the active site, but are unable to be converted to products.

Skills Test Problem

Food rots about 40 times more rapidly at 25 ^{0}C than at 4 ^{0}C. Estimate the overall activation energy for the processes responsible for decomposition. How fast would you expect the decomposition to be at -20 ^{0}C (freezer temperature)?

Solution

Use the Arrhenius equation in the form

$$\ln\frac{k_2}{k_1} = \frac{E_a}{R}\left(\frac{1}{T_1} - \frac{1}{T_2}\right)$$

$k_2 = 40k_1$, $T_2 = 298$ K and $T_1 = 277$K

$$\ln\frac{40k_1}{k_1} = \frac{E_a}{8.3245\ \text{Jmol}^{-1}\text{K}^{-1}}\left(\frac{1}{277} - \frac{1}{298}\right)$$

Solving for E_a gives $E_a = 1.21 \times 10^5$ J mol^{-1} = 121 kJ mol^{-1}.

At -20^{0}C (253 K), the rate is

$$\ln\frac{k_2}{k_1} = \frac{1.21\times10^5\ \text{Jmol}^{-1}}{8.3145\ \text{Jmol}^{-1}\text{K}^{-1}}\left(\frac{1}{277} - \frac{1}{253}\right)$$

$$= -4.98$$

or $\ln(k_1/k_2) = 4.98$ and $k_1/k_2 = 145$ or food rots 145 times slower at -20 ^{0}C.

Quiz A

1. For 2 NO + O_2 --> 2 NO_2 initial rate data are

[NO], M	0.010	0.030	0.010
$[O_2]$, M	0.010	0.020	0.020
Rate, M s^{-1}	2.5×10^{-9}	45.0×10^{-9}	5.0×10^{-9}

The rate is: rate = $k[NO]^x[O_2]^y$. What are the values for x and y?

2. For the elementary reaction 2 A --> 2 B + C, write a rate law in terms of [A], [B], and [C].
3. What is the first-order rate constant when the half-life is 3.0 sec?
4. For a second-order reaction, the first half-life is 10 min. The second half-life is

   a. 10 min b. 5 min c. 20 min d. 7.5 min e. none of these

5. Which of the following does NOT influence the rate of a reaction

   a. the value of ΔH
   b. the activation energy
   c. temperature of the reaction
   d. all three of these

6. For a first-order reaction, how does the <u>rate constant</u> for the reaction change when the concentration of reactants are doubled?

   a. the rate constant doubles
   b. the rate constant is reduced by 1/2
   c. cannot be determined
   d. the rate constant remains unchanged

7. For a given reaction, the rate = k[A][B][C]. What is the overall order of the reaction?
8. What is an activated complex in a reaction?
9. For a reaction, A--> products, a plot of 1/[A] vs. time yields a straight line. What is the order of the reaction?
10. For a given first order reaction, rate = k[A]. After 20 min. the concentration of [A] is 1/4 $[A]_0$. What is the half-life of the reaction?

Quiz B

1. For a given reaction, rate = $k[A][B]^2$ which of the following will increase the rate the most?

   a. doubling A
   b. doubling B
   c. tripling A
   d. halving B

2. The rate constant for a first-order reaction is 1.5 sec^{-1}. What is the half-life of the reaction?

3. The catalyst for a reaction
   a. changes the products formed
   b. requires less reactant
   c. is not changed in the reaction
   d. all of these
   e. none of these
4. For radioactive decay,
   a. the reaction is always fast
   b. the reaction is a first-order process
   c. the rate cannot be determined
   d. the order of the reaction depends on the isotope that is decaying
   e. none of the above
5. Using collision theory, explain why the rate of a reaction generally increases sharply with temperature.
6. Describe what is meant by a reaction mechanism.
7. The half-life of a first order reaction is 5.0 sec. What is the rate constant?
8. What are the units of k, the rate constant, for a third-order reaction? (Assume time is in seconds.)
9. A first-order reaction, A --> products, has a rate of reaction of 0.050 Ms^{-1} when [A] = 0.855 M. What is the rate constant, k, for this reaction?
10. For a given reaction, rate = $k[A][B]^2[C]^{-1}$. What is the overall order of the reaction?

Quiz C

True or False

1. An elementary step in a reaction mechanism is the fastest step in the reaction.
2. The rate law for an overall reaction is determined experimentally.
3. The half-life of a reaction is the time required for the reactant concentration to drop to half of its initial value.
4. The reaction order is determined by the number of components in the reaction.
5. Values of the exponents in a rate law are determined by the reaction mechanism.
6. The rate of a reaction stays the same as the reaction proceeds.
7. A catalyst increases the rate of a reaction by decreasing the activation energy.
8. For a first order reaction, the half-life is inversely proportional to the rate constant.
9. Enzymes are examples of biological catalysts.
10. The transition state in a reaction is at the minimum of the potential energy profile.

Self Test

Fill in the blank

1. The overall order for the rate law, rate = k[A][B] is ______________.
2. The half-life of a second-order reaction depends on ___________________________.
3. The rate of a reaction generally ____________ with temperature.
4. The __________________ can be determined if the values of the rate constants are known at different temperatures.
5. The _______________ is the high energy intermediate species resulting from favorable molecular collisions.

Multiple Choice

6. If the concentration of A is doubled in a reaction that is second-order in A, the reaction rate will
   a. stay the same
   b. double
   c. quadruple
   d. be reduced by half
   e. cannot be determined
7. A plot of 1/[A] vs t for a reaction yields a straight line. The reaction is
   a. zero order in [A]
   b. first order in [A]
   c. second order in [A]
   d. cannot be determined
8. A catalyst
   a. does not appear in the overall chemical equation
   b. decreases the activation energy
   c. is assumed to be unchanged in the reaction
   d. all of the above
   e. b and c
9. The rate determining step in a mechanism
   a. is first order
   b. is the slowest step
   c. must be reversible
   d. uses a catalyst
10. For a first-order reaction,
    a. the unit of the rate constant is 1/time
    b. the half-life does not depend on the initial concentration
    c. has a low activation energy

d. all of the above

e. two of the above

Problems/Short Answer

For Problems 11-13, refer to the mechanism below:

$A + B \rightarrow C$ slow

$C + B \rightarrow D$ fast

11. Which step is the rate determining step?
12. Which species is an intermediate, if any?
13. Write the rate law for the first step.
14. The rate law for the reaction $2A + B \rightarrow 2C + D$ was found to be rate = k[A][B]. Could the reaction occur in a single elementary step? Why?
15. The activation energy for a reaction is 120 kJ/mol. How many times greater is the rate constant for this reaction at 250 K than it is at 100 K?

CHAPTER 14

Chemical Equilibrium

This chapter will deal with reversible reactions, rather than reactions that go to completion. To deal with these reactions, we will use a quantity called the equilibrium constant and apply it to reactants and products in a reaction.

Chapter Objectives: You should be able to:

1. Describe the dynamic nature of equilibrium.
2. Write equilibrium constant expressions from the chemical equation.
3. Write equilibrium expressions in terms of pressures, K_p, for reactions involving gases.
4. Write equilibrium constant expressions for heterogeneous reactions.
5. Use the value of K_c or K_p to determine the magnitude of a reaction.
6. Use the reaction quotient to predict the direction of a reaction as it moves toward equilibrium.
7. State Le Chatelier's Principle in your own words and use this principle to describe what happens under changing conditions in a chemical system at equilibrium.
8. State how changes in pressure or volume affect equilibria involving gases.
9. State how temperature affects the equilibrium position.
10. Calculate the equilibrium constant from the amount of substances at equilibrium.
11. Calculate the amounts of substances at equilibrium from the equilibrium constants.

Chapter Summary

1. Dynamic equilibria

Equilibrium is the condition in which a reaction is reversible and there is no net change in the concentration of reactants and products. This condition does not mean that individual molecules are not undergoing changes, but only that the average concentrations are unchanged. Another way of stating this is that the rate of the forward and reverse reactions are the same.

2. Equilibrium constant expressions

The general form of the equilibrium expression for the following reversible reaction

aA + bB <--> cC + dD

$$K_c = \frac{[C]^c[D]^d}{[A]^a[B]^b}$$

When an equation is reversed, the value of K_c is inverted. We can see this is the case by using the rules for writing equilibrium expressions and writing the expressions for the two different reactions.

Exercise (text) 14.1A. If $[CO] = [Cl_2]$ at equilibrium for the reaction in Example 14.1 (text), is there just one possible value of $[COCl_2]$?

Solution: The equilibrium constant expression from Example 14.1 is $K_c = \dfrac{[COCl_2]}{[CO]\,[Cl_2]}$

If $[CO] = [Cl_2]$ then $[COCl_2] = K_c[CO]^2$ (or we could use $[Cl_2]^2$). This gives a number of possible values for $[COCl_2]$ depending on the values of [CO].

Example 14.1. K_c for the reaction $SO_2(g) + 1/2\ O_2(g) <--> SO_3(g)$ is 20.0 at 973 K. Calculate K_c for the reaction:

$2SO_3(g) <--> 2SO_2(g) + O_2(g)$.

Solution: This reaction is the reverse of the first reaction and the coefficients are multiplied by two. To write the equilibrium constant expression for the reverse reaction, take the inverse of the equilibrium constant, and to write the equilibrium constant expression for doubling the coefficients, square the equilibrium constant. To see this more clearly, write both equilibrium constant expressions:

For the first reaction:

$$K_c = \frac{[SO_3]}{[SO_2]\,[O_2]^{1/2}}$$

For the second reaction:

$$K_{c'} = \frac{[SO_2]^2[O_2]}{[SO_3]^2}$$

We see that $K_c' = 1/K_c^{\,2} = \dfrac{1}{(20.0)^2} = 2.50 \times 10^{-3}$

3. Equilibrium expressions in terms of pressures, K_p

$K_p = K_c(RT)^{\Delta n}$ where Δn is the sum of the coefficients of the gaseous products minus the sum of the coefficients of the gaseous reactants in the balanced equation.

Exercise (text) 14.3A Given $K_c = 1.8 \times 10^{-6}$ for the reaction $2\ NO(g) + O_2(g) \leftrightarrow 2\ NO_2(g)$ at 457 K, derive the value of K_p at 457 K for the reaction:

$NO_2(g) <--> NO(g) + 1/2\ O_2(g)$

Solution: First write K_c for the reaction and then convert this to K_p. $K_{c'}$ for the second reaction will be the inverse of the first K_c because the reaction is reversed. The coefficients are also multiplied by 1/2 and so the inverse of the equilibrium constant will be raised to the 1/2 power (the square root).

$K_c' = (1/K_c)^{1/2} = 0.75 \times 10^3 = 7.5 \times 10^2$

To convert to K_p:

$K_p = K_c(RT)^{\Delta n}$ and $\Delta n = 1.5 - 1 = 0.5$

$K_p = 7.5 \times 10^2 (0.08206 \times 457)^{1/2} = 4.6 \times 10^3$

4. Equilibrium constant expressions for heterogeneous reactions

In heterogeneous reactions, the concentration of substances that do not change in the reaction, such as pure solid or liquid phases, do not appear in the equilibrium constant expressions.

Exercise (text) 14.4A. Write the K_p expression for the water-gas reaction of Example 14.4.

Solution: The reaction is

$C(s) + H_2O(g) <--> CO(g) + H_2(g)$

The K_p expression will be similar to the K_c expression except that concentrations are expressed as partial pressures. As in Example 16.4, the concentration of solid C does not appear in the equilibrium expression. The equilibrium constant expression is:

$$K_p = \frac{(P_{H_2})(P_{CO})}{P_{H_2O}}$$

5. Using the value of K_c or K_p to determine the magnitude of a reaction

A very large numerical value of K_c or K_p indicates that a reaction goes to completion or very nearly so. A very small value indicates that the reaction does not occur to any significant extent.

Exercise (text) 14.5A. Refer to Example 14.1 and determine if we can assume that the reaction $CO(g) + Cl_2(g) --> COCl_2(g)$ goes essentially to completion at 395 °C. Explain your reasoning.

Solution: The equilibrium constant gives us the ratio of the concentration of product to the product of the reactant concentrations. If we assume the reactant concentrations are equal, then the product concentration is approximately 1000 times greater than the product of the reactant concentrations. This means the reaction goes essentially to completion. If we substitute values of 1.0 for the reactant concentrations, this result is very readily apparent. A similar result will be obtained with other values.

6. Reaction Quotient

For nonequlibrium conditions, the expression having the same form as the equilibrium constant is the reaction quotient, Q_c or Q_p. The reaction quotient allows us to predict the direction of the reaction as it goes to equilibrium. If $Q_c < K_c$ a net reaction proceeds in the forward direction, that is, from left to right. If $Q_c > K_c$, a net reaction proceeds in the reverse direction, that is, from right to left.

Example 14.2. What will be the amounts of reactants and products when equilibrium is established in a gaseous mixture that initially has 0.0100 mol H_2 and 0.100 mol HI in a 5.25-L volume at 698 K?

$H_2(g) + I_2(g) \leftrightarrow 2\,HI(g)$ $\quad K_c = 54.3$ at 698 K

Solution: If we look at the K_c expression, we realize that the volumes will cancel and we can work with moles. Also, since there is no I_2, one of the reactants, present initially, the reverse reaction must occur to establish equilibrium. To calculate the equilibrium concentrations:

	H_2+	I_2 <-->	2HI
Initial conc.	0.0100	0.00	0.100
Change	+x	+x	-2x
Equilibrium	0.01 + x	x	0.100-2x

$$K_c = \frac{(0.100-2x)^2}{(0.0100+x)(x)} = \frac{0.010-0.40x+4x^2}{0.0100x+x^2} = 54.3$$

Rearranging to a quadratic gives:

$50.3x^2 + 0.943x - 0.0100 = 0$

Solving the quadratic for x:

$x = 7.56 \times 10^{-3}$ mol

The equilibrium amounts are:

$H_2 = 0.0176$ mol $\quad I_2 = 7.56 \times 10^{-3}$ mol $\quad HI = 0.085$ mol

Most of the examples so far have worked with concentrations or amounts in moles. However, for gases we can also work with pressures.

Example 14.3. Ammonium hydrogen sulfide dissociates readily, even at room temperature. What is the total pressure of the gases in equilibrium with $NH_4HS(s)$ at 25 °C?

$NH_4HS(s) \leftrightarrow NH_3(g) + H_2S(g)$ $\quad K_p = 0.108$ at 25 °C

Solution: Assuming that the source of the two gases is from dissociation only, the pressures of the two gases must be equal since the moles will be equal. Solve for the pressure of either gas and the sum of the two pressures will be the total pressure.

$K_p = P_{NH_3} \times P_{H_2S}$

$= x^2 = 0.108$

$X = (0.108)^{1/2} = 0.329 \text{ atm} = P_{NH_3} = P_{H_2S}$

and the total pressure is the sum of the two pressures:

$P_t = 0.658 \text{ atm}$

7. Le Chatelier's Principle

When a change in a system at equilibrium occurs (removal of product or increase in reactant concentration, for example), the system readjusts to partially offset the change. For example, if a system A + B <--> C is at equilibrium and we begin removing C, the system will adjust so that more A and B react to produce C.

Exercise (text) 14.7A. What should be the effect of each of the following changes on a constant-volume equilibrium mixture of N_2, H_2, and NH_3?

$N_2(g) + 3\ H_2(g) \text{ <--> } 2\ NH_3(g)$

a. adding $H_2(g)$

b. removing $N_2(g)$

c. removing $NH_3(g)$.

Solution: According to Le Chatelier's principle, a system at equilibrium will change when the conditions are changed to partially offset the effects of the new conditions.

- **a.** Addition of reactant, H_2 will increase the amount of product and increase the forward reaction.
- **b.** Removal of reactant will increase the reverse reaction. At the new equilibrium, there will be less NH_3 and also less N_2, but the concentration of N_2 will not be as small as the concentration immediately after removal. Some will re-form from the reverse reaction.
- **c.** Removal of product. This will increase the forward reaction so that some of the NH_3 removed will be replaced. There will still be less NH_3 at the new equilibrium conditions.

8. Changes in pressure or volume affect equilibria involving gases

For an equilibrium mixture of gases, when the volume is decreased by increasing the pressure, the equilibrium shifts in the direction producing a smaller number of moles of gas. When the volume of an equilibrium gas mixture is increased by decreasing the pressure, the equilibrium shifts in the direction producing a larger number of moles of gas.

Example 14.4. How is the equilibrium amount of HI(g) in the reaction

$H_2(g) + I_2(g) \text{ <--> } 2\ HI(g)$ changed by changing the reaction volume?

Solution: We have just stated how the equilibrium shifts for changes in pressure, but in this case the number of moles of gas remains constant. The equilibrium amounts will also remain constant and will not shift. If we examine the equilibrium constant expression,

$$K_p = \frac{(P_{H_2})(P_{I_2})}{P_{HI}^2}$$

we see that reducing the volume by 1/2 will increase all pressures by a factor of two. This will increase the numerator by a factor of four and the denominator by a factor of four. Therefore, there is no shift in the equilibrium amounts.

9. Temperature and equilibrium position

Raising the temperature of an equilibrium mixture shifts the equilibrium in the direction of the endothermic reaction; lowering the temperature shifts the equilibrium in the direction of the exothermic reaction. Writing the reactions with "heat" as a reactant or product makes this relationship easier to remember.

Example 14.5. Is the conversion of $SO_2(g)$ to $SO_3(g)$ more nearly complete at high or low temperatures?

$$2\ SO_2(g) + O_2(g) \text{ <--> } 2\ SO_3(g) \qquad \Delta H^\circ = -198 \text{ kJ}$$

Solution: ΔH° is for the forward reaction. Since ΔH° is negative, the forward reaction is exothermic. In order for the reaction to be nearly complete, we want to favor the forward reaction. Lowering the temperature will favor the exothermic, forward reaction and the reaction will be more nearly complete at low temperatures.

What is the effect of adding a catalyst to a system at equilibrium?

A catalyst will change the rate of reaction but will not change the position of equilibrium.

10. Calculating the equilibrium constant from the amount of substances at equilibrium

To calculate the equilibrium constant, write the appropriate equilibrium expression and substitute in the equilibrium concentrations in appropriate units.

Exercise (text) 14.11B. A 1.00-kg sample of $Sb_2S_3(s)$ and 10.0 g $H_2(g)$ are allowed to react in a 25.0-L container at 713 K. At equilibrium, 72.6 g $H_2S(g)$ is present. What is the value of K_p at 713 K for the reaction:

$$SbS_3(s) + 3\ H_2(g) \text{ <--> } 2\ Sb(s) + 3\ H_2S(g)?$$

Solution: Write the equilibrium expression for K_p (or K_c and later convert) and determine the concentration of the species involved in the equilibrium.

$$K_c = \frac{[H_2S(g)]^3}{[H_2(g)]^3}$$

We are given the mass of $H_2S(g)$ at equilibrium. We only need to determine the equilibrium amount of $H_2(g)$ since the solids will not enter into the equilibrium constant expression. The amount of H_2 consumed is determined from the reaction stoichiometry.
Calculate the amount of H_2 from the reaction by calculating the initial amount in moles and the amount in moles used to form H_2S. The initial amount of H_2 is:

amount H_2 = 10.0 g x (1 mol/2.016 g) = 4.96 mol

The equilibrium amount of H_2S are:

amount H_2S = 72.6 g x (1mol/34.08 g) = 2.13 mol

$SbS_3(s) + 3\ H_2(g) \longleftrightarrow 2\ Sb(s) + 3\ H_2S(g)$

	$H_2(g)$	$H_2S(g)$
Initial amount	4.96 mol	0
Change	-2.13 mol	+2.13 mol
Equil. amount	2.83 mol	2.13 mol

From the reaction stoichiometry, to form 2.13 mol of H_2S requires 2.13 mol of H_2. Therefore, the equilibrium amount of H_2 must be 4.96 mol - 2.13 mol = 2.83 mol. We now have sufficient information to calculate K_c:

$$K_c = \frac{[H_2S(g)]^3}{[H_2(g)]^3} = \frac{\left(\frac{2.13\ \text{mol}}{25.0\ \text{L}}\right)^3}{\left(\frac{2.83\ \text{mol}}{25.0\ \text{L}}\right)^3} = 0.426.$$

To convert to K_p:

$K_p = K_c(RT)^{\Delta n}$

but $\Delta n = 0$ since there is no change in the number of moles of gas and $(RT)^{\Delta n} = 1$.
Therefore $K_p = K_c = 0.426$.

11. Determining equilibrium quantities from the equilibrium constant

Example 14.6. What will be the mole fraction of NO(g) at equilibrium if an equimolar mixture of $N_2(g)$ and $O_2(g)$ is brought to equilibrium at 2500 K?

$N_2(g) + O_2(g) \longleftrightarrow 2\ NO(g)$ $K_c = 2.1 \times 10^{-3}$ at 2500 K

Solution: We can assume any amount of N_2 and O_2 so long as they are equal. Also, note that the volume of the system doesn't matter because the volume term will cancel in the equilibrium expression. Write the amounts in moles, and the changes to calculate the equilibrium amount of each substance. (We could assume a volume of 1.0 L and work with concentrations instead of amounts if we chose.) If we assume 1.00 mole of N_2 and O_2 then:

	$N_2(g)$ +	$O_2(g)$ <-->	2 NO(g)
Initial amount	1.00 mol	1.00 mol	0.00
Change	-x	-x	+2x
Equilibrium amount	1.00 - x	1.00 - x	2x

Writing the K_c expression:

$K_c = \frac{(2x)^2}{(1-x)^2}$ if we assume that 1- x is approximately = 1, we can solve for K_c:

$K_c = (2x)^2 = 2.1 \times 10^{-3}$ and solving for x we obtain:

x = 0.023 mol which is small compared to 1.00. The amount of NO is 2x = 0.046 mol. The mole fraction is moles NO/moles total and

$X_{NO} = 0.046$ mol NO/(2.0 mol total) = 0.023

Remember the amount of N_2 and O_2 is still approximately 1.00 mole, but the total is two moles.

In the above example, we assumed that the change in the two reactants was small relative to the initial concentrations and subsequent calculations showed this to be a reasonable assumption. Often, however, we cannot make this assumption and then we must use the quadratic equation to solve for the concentrations.

Exercise (text) 14.13B. How many moles of NO(g) will form at 2500 K when equilibrium is established in a mixture that initially has 0.78 mol N_2 and 0.21 mol O_2 ?

$N_2(g) + O_2(g)$ <--> 2 NO(g) $\qquad K_c = 2.1 \times 10^{-3}$ at 2500 K

Solution: Solve this in the same way we solved the previous example, except now our assumption that the change in concentration with respect to the initial concentration is small will probably not be valid. Consequently we will have to solve a quadratic equation. Proceeding as before:

	$N_2(g)$ +	$O_2(g)$ <-->	2 NO(g)
Initial amount	0.78 mol	0.21 mol	0.00
Change	-x	-x	+2x
Equilibrium amount	0.78 - x	0.21 - x	2x

$$K_c = \frac{(2x)^2}{(0.78-x)(0.21-x)} = 2.1 \times 10^{-3} = \frac{4x^2}{0.164 - 0.99x + x^2}$$

Rearranging the equation in the form of a quadratic gives:

$4x^2 + 2.1 \times 10^{-3}x - 3.44 \times 10^{-4} = 0$

Solving the quadratic equation:

$$x = \frac{-b \pm \sqrt{b^2 - 4ac}}{2a} = \frac{-2.1x10^{-3} + \sqrt{(2.1x10^{-3})^2 - 4(4)(-3.44x10^{-4})}}{2(4)}$$

$= 9.0 \times 10^{-3}$

The amount in moles of NO that will form will be 2x or 1.8×10^{-2} mol.

Quiz A

1. Write the equilibrium constant, K_c, expression for the reaction

   $2\ Fe(s) + 3\ Cl_2(g)$ <--> $2\ FeCl_3\ (s)$

2. For the reaction in a 10.0L container:

   $CO(g) + 2\ H_2(g)$ <--> $CH_3OH(g)$ $K_c = 14.5$ at 500 K

   Initially 0.10 mol of CO (g), 0.20 mol of H_2, and 6.0 mol of CH_3OH are present. When equilibrium is reached, will the concentration of each reactant or product increase or decrease?

3. For an equilibrium where the forward reaction is exothermic, will increasing the temperature alone favor formation of reactants or products?

4. For the reaction:

   $2\ Fe(s) + 3\ Cl_2(g)$ <--> $2\ FeCl_3\ (s)$

   what is the effect of adding solid $FeCl_3$ on the concentration of Cl_2 present?

5. Adding a catalyst to an equilibrium mixture will

   a. shift the equilibrium towards products

   b. increase the rate of the forward reaction only

   c. shift the equilibrium towards reactants

   d. cause no change in the equilibrium concentrations of reactants and products

   e. none of these

6. Define equilibrium.

7. If the coefficients in a balanced chemical equation are multiplied by two, the equilibrium constant will be

   a. the same

   b. the square root

   c. divided by two

   d. the square.

8. Write the expression relating K_p to K_c for a reaction involving gases.

9. A very large numerical value for K_c indicates
   a. the reaction goes to completion
   b. the reaction does not occur
   c. the reaction is very rapid
   d. the reaction is very slow
   e. none of these
10. What is meant by a heterogeneous equilibrium?

Quiz B

1. For the reaction $N_2(g) + 3 H_2(g) <--> 2 NH_3(g)$, write the expression for K_c.
2. For the above reaction, write the expression for K_p in terms of K_c.
3. For the reaction $N_2(g) + 3 H_2(g) <--> 2 NH_3(g)$, the initial conditions are 0.2 mol $N_2(g)$, 0.00 mol $H_2(g)$ and 0.01 mol $NH_3(g)$. Will the concentration of reactants and products increase or decrease when equilibrium is established?
4. A very small value of K_c indicates (a) the reactions are very rapid (b) the forward reaction does not occur to any significant extent (c) the reverse reaction does not occur to any significant extent (d) the temperature is too high (e) all of the above
5. If Q, the reaction quotient, is greater than K_c, which direction will the reaction proceed to obtain equilibrium?
6. For the reaction $CaCO_3(s) <--> CaO(s) + CO_2(g)$, what will be the effect of removing some of the CaO(s) on the equilibrium?
7. For the reaction

   $N_2(g) + 3 H_2(g) <--> 2 NH_3(g)$
   what will be the effect of decreasing the volume on the equilibrium position?
8. For an equilibrium where the forward reaction is endothermic, what will be the effect of increasing temperature?
9. For the reaction $CO(g) + H_2O(g) <--> CO_2(g) + H_2O(g)$, K_c =9.0 at 698 K. What is the value of K_c for the reaction $2 CO_2(g) + 2 H_2O(g) <--> 2 CO(g) + 2 H_2O(g)$?
10. Define homogeneous equilibrium.

Quiz C

True or False

1. For a reaction at equilibrium, the rates of the forward and reverse reactions are equal.
2. A catalyst shifts the equilibrium toward products.
3. A change in temperature always changes the equilibrium constant in a process involving at least one gaseous reactant or product..
4. If a reactant is added to an equilibrium mixture, the reaction will shift to the left.

5. An equilibrium is reached when the concentrations of reactants and products have reached a constant value.
6. A heterogeneous equilibrium is when the forward reaction has a different rate law from the reverse reaction.
7. For an exothermic reaction, K_c increases with temperature.
8. When a catalyst is added to a system at equilibrium, a new lower energy pathway is established.

Self Test

Fill in the blank

1. Multiplying the coefficients of a reaction by two ________________ the equilibrium constant.
2. The equilibrium constant expression for the reverse of a reaction is ________________ of the equilibrium constant in the forward direction.
3. In a heterogeneous equilibrium, increasing the amount of pure solid will ________________ the point of equilibrium.
4. The ________________ compared to the equilibrium constant predicts the direction the reaction will move to achieve equilibrium.
5. ________________________ states that when a system is at equilibrium, if a change is made to the equilibrium conditions the system will change to establish a new equilibrium.
6. For an equilibrium involving gases, an increase in pressure will result in a net reaction that ____________________ the total concentration of gas molecules.

Problems/Short Answer

Refer to the reaction below for problems 7-12
For the reaction

$Cu_2S\ (l) + O_2(g) \leftrightarrow 2Cu(l) + SO_2\ (g)\ \Delta H\ –250\ kJ$

7. Write the equilibrium constant expression.
8. What is the effect of increasing temperature?
9. What is the effect of increasing pressure?
10. What is the effect of adding Cu(l)?
11. What is the effect of adding $O_2(g)$?
12. What is the equilibrium constant expression for the reverse reaction?
13. For the reaction $PCl_5(g) \leftrightarrow PCl_3\ (g) + Cl_2(g)$ $K_c = 0.24$ at 300 °C.
14. What is the value for K_p?
15. Referring to the above reaction in Prob. 13, is the following mixture at equilibrium?

$[PCl_5] = 1.2$ mol/L $[PCl_3] = 0.5$ mol/L $[Cl_2] = 2.3$ mol/L

16. Calculate all equilibrium concentrations from the information in Prob. 13 and 14.

CHAPTER 15

Acids, Bases and Acid-Base Equilibria

In this chapter we will examine the factors that affect acid and base strength. We will describe the quantitative measure of acids and bases, pH, and calculate pH of solutions. Neutralization of acids and bases will be discussed.

Chapter Objectives: You should be able to:

1. Define Bronsted-Lowry acids and bases.
2. Determine the favored direction of reaction for a conjugate acid/base pair.
3. State the relationship between bond strength and strength of acids and how electronegativity relates to acid strength for oxoacids.
4. Write the ion product of water.
5. Define and know how to calculate pH and pOH.
6. Calculate pH values for weak acids and bases.
7. Determine the K_a or K_b of a weak acid or weak base from the concentration of species in an equilibrium.
8. Define polyprotic acids and calculate the pH for a solution containing a polyprotic acid.
9. Describe how ions can act as acids and bases.
10. State the relationship between pK_a, pK_b and pK_w
11. Describe the common ion effect.
12. Define a buffer solution and state how it works.
13. Define a pH indicator and state how it is used.
14. Define equivalence point.
15. Describe the qualitative differences in the titration curve of a strong acid-strong base and weak acid-strong base.
16. Define Lewis acids and bases.

Chapter Summary

1. Bronsted-Lowry theory of acids and bases

Bronsted-Lowry theory describes acids as proton donors and bases as proton acceptors. As in the case of electron transfer, if there is a proton donor there must also be a proton acceptor. An acid will donate a proton; the remaining ion is able to accept a proton and is the conjugate base of the acid. Similarly, a base will have a conjugate acid, which is the base after it has accepted a proton.

A substance that can act either as an acid or base is amphoprotic. Some examples of amphiprotic substances are water, and many oxides such as Al_2O_3, PbO, ZnO, Cr_2O_3, etc.

Example 15.1. Identify the Bronsted-Lowry acids and bases and their conjugates in each of the following ionizations:

a. $HS^- + H_2O \leftrightarrow H_2S + OH^-$

b. $HNO_3 + H_2PO_4^- \leftrightarrow H_3PO_4 + NO_3^-$

Solution: In each case we must identify the proton donors and acceptors. In

a. H_2O is the acid and OH^- its conjugate base. HS^- is a base and H_2S is the conjugate acid. In the first case H_2O donated a proton to become OH^- which will accept a proton (base) in the reverse reaction.

b. HNO_3 donates a proton (acid) to become NO_3^- (the conjugate base). $H_2PO_4^-$ accepts a proton (base) to become H_3PO_4 (the conjugate acid).

2. Reaction of a conjugate acid/base pair

In a conjugate acid/base pair, the reaction is favored in the direction from the stronger to weaker member of the pair. A strong acid/weak base pair will favor reaction from the strong acid to the weak base. For example, the strong acid/weak base pair, HI/I^- will react in solution

$HI + H_2O \leftrightarrow H_3O^+ + I^-$

Since I^- is the weak base, the reaction will favor the formation of I^-.

3. Bond strength, electronegativity and strength of acids

For binary acids, the stronger the bond (higher bond energy) between hydrogen and the other element, the weaker the acid. An acid must dissociate in order to react in solution and if the bond energy is high, little dissociation will take place and the compound will be a weak acid.

For oxoacids, such as HOI and HOCl, where a nonmetallic element, E, is bonded to oxygen, the acid strength increases as the electronegativity of the element E increases. The more electronegative the element, the greater the tendency to draw electrons away from the O-H bond and the weaker the bond will be. This will lead to easier dissociation of the H^+ ion and a stronger acid.

Exercise (text) 15.2A. Select the stronger acid in each of the following pairs:

a. H_2S or H_2Te

b. $CH_3CH_2CH_2CHBrCOOH$ or $ClCH_2CH_2CH_2CH_2COOH$

Solution:

a. H_2Te should be the stronger acid because as anion radius decreases, acid strength decreases. Moving down the periodic table, S^{2-} is a smaller anion than Te^{2-}.

b. $CH_3CH_2CH_2CH_2BrCOOH$ is the stronger acid because the Br attached to the C will have electron withdrawing effects and weaken the OH bond of the acid, making it a stronger acid.

4. Ionization of water

Water self-ionizes to give H_3O^+ and OH^-, but the extent of ionization is small. The ion product of water is $K_w = [H_3O^+][OH^-] = 1.0 \times 10^{-14}$ at 25 °C.

5. pH and pOH

Definitions of pH and pOH are:

$pH = -\log[H_3O^+]$ and $pOH = -\log[OH^-]$

We can define pK_w as $-\log K_w$ and $pK_w = pH + pOH = 14.00$.

Example 15.2. What is the pH of a solution prepared by dissolving 0.0155 mol $Ba(OH)_2$ in water to give 735 mL of aqueous solution? Assume that the $Ba(OH)_2$ is completely dissociated.

Solution: Since the dissociation of $Ba(OH)_2$ will produce hydroxide ions, it is easier to calculate pOH and then find the pH. First write the equation for the reaction:

$Ba(OH)_2(aq) \rightarrow Ba^{2+}(aq) + 2\,OH^-(aq)$

Since we have 0.0155 mol $Ba(OH)_2$, we will have

$$\frac{2 \text{ mol } OH^-}{\text{mol } Ba(OH)_2} \times 0.0155 \text{ mol } Ba(OH)_2 = 0.0310 \text{ mol of } OH^-.$$

We must now calculate the molarity which is 0.0310 mol/0.735 L = 0.0422 M. To calculate pOH,

$pOH = -\log 0.0422 = 1.375$, and to calculate pH,

$pH = 14 - pOH = 14 - 1.375 = 12.63$

A good check is to notice that this is a very basic solution and $Ba(OH)_2$ is a strong base, so this makes sense.

Example 15.3. Is a solution that is 1.0×10^{-8} M NaOH acidic, basic or neutral? Explain.

Solution: Because NaOH is a strong base, the solution will be very slightly basic. The self ionization of water produces more OH^- than the base itself. From the two sources, however, the OH^- will be greater than 10^{-7} and the solution will be basic.

6. pH values for weak acids and bases

Weak acids and weak bases are not completely ionized. To calculate pH values for solution of weak acids and weak bases, we need to use the appropriate ionization constant expressions, K_a and K_b.

Example 15.4. Determine the pH of 0.250 M C_6H_5COOH (benzoic acid). Obtain the K_a value from Appendix C.

Solution: First write the equilibrium expression. Using K_a we can obtain the concentration of H^+ and the pH of the solution.

$C_6H_5COOH + H_2O \leftrightarrow H_3O^+ + C_6H_5COO^-$ $\quad K_a = 6.3 \times 10^{-5}$

We can neglect the self ionization of water since K_a is so much larger than K_w.

	$C_6H_5COOH + H_2O \leftrightarrow$	H_3O^+ +	$C_6H_5COO^-$
Init. conc.	0.250	--	--
Change	-x	+x	+x
Equil. conc.	0.250 - x	x	x

$$K_a = \frac{[H_3O^+][C_6H_5COO^-]}{[C_6H_5COOH]} = \frac{x^2}{0.250 - x} = 6.3 \times 10^{-5}$$

We can try the approximation that x is small relative to 0.250 M and solve the equation without using the quadratic equation.

$x^2 = (6.3 \times 10^{-5})(0.250) = 1.6 \times 10^{-5}$

$x = 4.0 \times 10^{-3}$ M (This is less than 2% of 0.250 M so the approximation is adequate.)

$[H_3O^+] = 4.0 \times 10^{-3}$ M and pH = -log $[H_3O^+]$ = - log $(4.0 \times 10^{-3}) = 2.40$

Example 15.5. Calculate the pH of 0.0010 M $NH_3(aq)$. Show that even though the "5% rule" for the simplifying assumption fails to two significant figures, the same result is obtained with and without the assumption.

Solution: First, we write the ionization reaction for NH_3.

	$NH_3 + H_2O \leftrightarrow$	NH_4^+ +	OH^-	$K_b = 1.8 \times 10^{-5}$
Init. conc.	0.001	--	--	
Change	-x	+x	+x	
Equil. conc.	0.001 - x	x	x	

$$K_b = \frac{[NH_4^+][OH^-]}{[NH_3]} = \frac{x^2}{0.001 - x} = 1.8 \times 10^{-5}$$

Using the simplifying assumption that x is < 0.001 we get

$x^2 = 1.8 \times 10^{-8}$ and $x = 1.34 \times 10^{-4}$ M $\qquad$ pOH = -log $(1.34 \times 10^{-4}) = 3.87$

pH = 14.00 - 3.87 = 10.13

Since 1.34×10^{-4} is about 13% of the value of the concentration, the simplifying assumption is not valid and we should solve the quadratic equation. Writing the quadratic we get

$x^2 + (1.8 \times 10^{-5})x - 1.8 \times 10^{-8} = 0$

and

$$x = \frac{-b \pm \sqrt{b^2 - 4ac}}{2a} = \frac{-1.8 \times 10^{-5} \pm \sqrt{(1.8 \times 10^{-5})^2 - 4(-1.8 \times 10^{-8})}}{2} = 1.25 \times 10^{-4}\ M$$

$pOH = -\log(1.25 \times 10^{-4}) = 3.90$

$pH = 14.00 - 3.90 = 10.10$

We see that to two significant figures, our results are the same whether the simplifying assumption is used or not.

7. Determining the K_a or K_b of a weak acid or weak base from the concentration of species in an equilibrium

This is just calculating the equilibrium constant as we did earlier To obtain the concentrations, we may need to work with pH or pOH.

Example 15.6. Suppose you discovered a new acid, HZ, and found that the pH of a 0.0100 M solution is 3.12. What is K_a and pK_a for HZ?

$HZ + H_2O \quad \text{<-->} \quad H_3O^+ + Z^-$

Solution: First, calculate the equilibrium concentration of H_3O^+ using the pH.

$pH = 3.12 = -\log[H_3O^+]$

$[H_3O^+] = 7.59 \times 10^{-4}\ M$

Using the ionization equation

	HZ	+ H_2O <-->	H_3O^+	+ Z^-
Init. conc.	0.0100 M		--	--
Change	-7.59 x 10^{-4} M		+ 7.59 x 10^{-4} M	+ 7.59 x 10^{-4} M
Equil. conc.	0.0100 M -7.59 x 10^{-4} M		7.59 x 10^{-4} M	7.59 x 10^{-4} M

$$K_a = \frac{[H_3O^+][Z^-]}{[HZ]} = \frac{(7.59 \times 10^{-4})^2}{9.2 \times 10^{-3}} = 6.3 \times 10^{-5}$$

$pK_a = -\log(6.3 \times 10^{-5}) = 4.20$

Example 15.7. Which solution is more basic, 0.025 M NH_3 or 0.030 M methylamine, CH_3NH_2 ?

Solution: Methylamine will be more basic because it has a much larger K_b and therefore will be more ionized. Since the ionization reaction produces OH^-, the more ionized species will be more basic.

8. Polyprotic acids

Polyprotic acids are acids that have more than one ionizable hydrogen atom per molecule.

Example 15.8. A solution used in cleaning boilers is 5% H_3PO_4 by mass and has a pH = 1.2. What are the concentrations of dihydrogen phosphate ion, $H_2PO_4^-$, and hydrogen phosphate ion, HPO_4^{2-} ?

Solution: We can assume that the concentration of dihydrogen phosphate is essentially the same as the $[H_3O^+]$ = 0.060 M since all of the $[H_3O^+]$ came from the ionization of H_3PO_4 and one mole of $H_2PO_4^-$ is formed for each H_3PO_4 ionized. We can calculate the concentration of the hydrogen phosphate ion using the ionization of the dihydrogen phosphate:

	$H_2PO_4^-$ $+H_2O$ <-->	HPO_4^{-2} +	H_3O^+	$K_a = 6.3 \times 10^{-8}$
Init conc.	0.06 M	-	0.06 M	
Change	-x	+x	+x	
Equil. conc.	0.06 M - x	x	0.06 M + x	

$$K_a = \frac{[H_3O^+][HPO_4^{-2}]}{[H_2PO_4^-]} = \frac{(0.06 + x)(x)}{0.06 - x} = 6.3 \times 10^{-8}$$

If we approximate 0.06 + x ~ 0.06 and 0.06 - x ~ 0.06 then the equation becomes $K_a = x = 6.3 \times 10^{-8}$ M = $[HPO_4^{2-}]$. The amount is due to the second ionization which is very small and the concentration of H_3O^+ results almost entirely from the first ionization.

Example 15.9. What is the approximate pH of 8.5×10^{-4} M H_2SO_4?

Solution: This is a case of low concentration where both ionizations go essentially to completion. The concentration of H_3O^+ is $2 \times 8.5 \times 10^{-4}$ since each mole of H_2SO_4 produces two mole of H_3O^+. $[H_3O^+] = 1.7 \times 10^{-3}$ M and pH = $-\log(1.7 \times 10^{-3}) = 2.77$.

9. Ions as acids and bases

Ions can act as acids or bases through hydrolysis reactions. Some generalizations about hydrolysis reactions are: (1) Salts of strong acids and strong bases form neutral solutions (e.g. NaCl). (2) Salts of weak acids and strong bases form basic solutions (sodium acetate). (3) Salts of strong acids and weak bases form acidic solutions (NH_4Cl). (4) Salts of weak acids and weak bases form solutions that are acidic, neutral, or basic. Give examples of each type of reaction.

Exercise (text) 15.13A. Indicate whether you expect each of the following solutions to be acidic, basic or neutral:

a. $NaNO_3(aq)$ **b.** $CH_3CH_2CH_2COOK$

Solution:

a. $NaNO_3$ is the salt of a strong acid, HNO_3. A strong acid is completely dissociated. Similarly, NaOH is a strong base. NO_3^- in solution does not hydrolyze and the solution will be neutral.

b. $CH_3CH_2CH_2COOK$ is the salt of a weak acid. The resulting solution will be basic. $CH_3CH_2CH_2COOK$ will hydrolyze in solution and form $CH_3CH_2CH_2COO^-$, which will react with H_3O^+ from water, reducing the amount of H_3O^+ and causing the resulting solution to be basic. K^+ can potentially react with OH^-, but KOH is a strong base and will be completely dissociated.

10. Relationship between pK_a, pK_b and pK_w.

$K_a \times K_b = K_w$ and $pK_a + pK_b = pK_w = 14.00$ (at 25 °C)

Example 15.20. What molarity CH_3COONa (aq) solution has a pH = 9.10?

Solution: To calculate the molarity, we need to find the necessary concentration of salt. We can write the ionization reaction and use K_b to calculate the concentration.

$$CH_3COO^- + H_2O \leftrightarrow CH_3COOH + OH^-$$

First we can calculate the $[OH^-]$ from the pH.

$$pOH = 14.00 - pH = 14.00 - 9.10 = 4.90 = -\log[OH^-]$$

$$[OH^-] = 1.26 \times 10^{-5}$$

We can assume that the ionization of water does not contribute to the concentration. This is also the $[CH_3COOH]$, since the source of OH^- is from the hydrolysis reaction. Using K_b expression:

$$K_b = 5.6 \times 10^{-10} = \frac{[CH_3COOH][OH^-]}{[CH_3COO^-]} = \frac{(1.26 \times 10^{-5})(1.26 \times 10^{-5})}{x}$$

Solving for x:

$$x = (1.59 \times 10^{-10})/(5.6 \times 10^{-10}) = 0.28 \text{ M}$$

This is the equilibrium concentration of $[CH_3COO^-]$. To get the total concentration, we should add the protonated form, but this is small compared to the salt concentration.

11. Common ion effect

The common ion effect is the suppression of the ionization of a weak acid or a weak base by the presence of a common ion from a strong electrolyte. This is essentially another application of Le Chatelier's principle. For example, $NH_3 + H_2O <--> NH_4^+ + OH^-$ Addition of NH_4^+ as a salt such as NH_4Cl will shift the reaction to the left.

Exercise (text) 15.16A. Calculate the pH of a solution that is 0.15 M NH_3 and 0.35 M NH_4NO_3.

$NH_3 + H_2O <--> NH_4^+ + OH^-$ $\qquad K_b = 1.8 \times 10^{-5}$

Solution: Using the initial concentrations given and writing the changes in concentration:

	$NH_3 + H_2O <-->$	$NH_4^+ +$	OH^-
Init. conc.	0.15 M	0.35 M	
Change	-x	+x	+x
Equil conc.	0.15 M - x	0.35 M + x	x

$$K_b = \frac{[NH_4^+][OH^-]}{[NH_3]} = \frac{(0.35 + x)(x)}{(0.15 - x)} = 1.8 \times 10^{-5}$$

If we assume that x is small relative to the initial concentrations, then:

$(0.35)x/(0.15) = 1.8 \times 10^{-5}$

$x = 7.7 \times 10^{-6}$ M and $pOH = -\log(7.7 \times 10^{-6}) = 5.1$

$pH = 14.0 - pOH = 14.0 - 5.1 = 8.9$

12. Buffers

A buffer solution is one that changes pH only slightly when small amounts of a strong acid or a strong base are added. A buffer solution requires two components at appreciable and comparable concentrations, one to neutralize acids and the other to neutralize bases. The two components, however, cannot neutralize each other. This rules out strong acids and strong bases. Most buffer solutions are mixtures of a weak acid and its conjugate base or a weak base and its conjugate acid.

The buffer capacity is the amount of acid or base a buffer solution can neutralize and maintain an essentially constant pH.

Exercise (text) 15.17A. What is the final pH if 0.03 mol HCl is added to 0.500 L of a 0.24 M NH_3-0.20 M NH_4Cl buffer solution?

Solution: In this case, we are adding a strong acid which will neutralize the OH^-. Enough OH^- will be formed to react with the HCl. We can write this by writing the neutralization reaction and using a negative sign for the OH^- addition to indicate that this is effectively being removed from the solution. We also need to convert to molar concentration for the calculations.

	$NH_3 + H_2O$ <-->	NH_4^+ +	OH^-
Init. conc.	0.24 M	0.20 M	
Addition			-0.060 M
Change	- 0.060 M	+ 0.060 M	+ 0.060 M
After neut.	0.24 M - 0.060 M	0.20 M + 0.060 M	0

To calculate the pH, we can use the Henderson-Hasselbach equation and the relationship between pK_a and pK_b.

$$pH = pK_a + \log\frac{[\text{conjugate base}]}{[\text{weak acid}]} = pK_a + \log\frac{[NH_3]}{[NH_4^+]} = pK_a + \log\frac{(0.18)}{(0.26)}$$

$$pH = 9.26 + (-0.16) = 9.10$$

The pH is lower, as expected, since a strong acid was added to the solution. However, because of the buffer effect, the pH does not change very much.

Exercise (text) 15.18B. What mass of NH_4Cl must be present in 0.250 L of 0.150 M NH_3 to produce a buffer solution with pH = 9.05?

Solution: To determine the mass, we need to calculate the amount of NH_4^+ required. Essentially all of the NH_4^+ will come from the NH_4Cl. Using the equation as in the previous example:

$$pH = pK_a + \log\frac{[\text{conjugate base}]}{[\text{weak acid}]} = pK_a + \log\frac{[NH_3]}{[NH_4^+]}$$

$$pH - pK_a = \log\frac{[NH_3]}{[NH_4^+]} = 9.05 - 9.26 = -0.21$$

$$10^{-0.21} = 0.62 = \frac{[NH_3]}{[NH_4^+]} = (0.150)/[NH_4^+]$$

$$[NH_4^+] = (0.150)/(0.62) = 0.24\ M = [NH_4Cl]$$

Mass of NH_4Cl = 0.24 M x 0.250 L x 53.49 g/mol = 3.2 g

13. pH Indicators

A pH indicator is a substance added to a titration mixture that changes color at the equivalence point.

14. Equivalence point

The equivalence point in a titration is the point at which the amount of acid and base are in stoichiometric proportions, with neither in excess.

The endpoint of a titration is the point in the titration where the indicator changes color. A properly chosen indicator will have an endpoint that corresponds closely to the equivalence point.

Exercise (text) 15.20A. For the titration described in Example (text) 15.20, determine the pH after the addition of the following volumes of 0.500 M NaOH:

a. 19.90 mL

b. 19.99 mL

c. 20.01 mL

d. 20.10 mL

Solution: First we can write the neutralization reaction and calculate the amount in moles of H_3O^+ present. Using the volumes and molarity of NaOH we can calculate how much is neutralized or how much OH^- is in excess.

From Example 15.19, we know that the H_3O^+ is 10 mmol.

a. 19.90 mL of base added

	H_3O^+ +	OH^- <--> $2H_2O$
Init. amounts	10.0 mmol	
Add, mmol	9.95	
changes	-9.95	-9.95
after rxn	0.05	~0

$[H_3O^+] = 0.05 \text{ mmol}/(20.0 \text{ mL} + 19.9 \text{ mL}) = 1.3 \times 10^{-3}$ M

$pH = -\log(1.3 \times 10^{-3}) = 2.9$

b. 19.99 mL of base added. Using the above approach, $[H_3O^+] = 0.005 \text{ mmol}/(39.99 \text{ mL}) = 1.3 \times 10^{-4}$ M

$pH = -\log(1.3 \times 10^{-4}) = 3.9$

c. 20.01 mL of base added. The mmol of OH^- = 0.5 x 20.01 = 10.005 In this case there will be an excess of OH^- after neutralization.

	H_3O^+ +	OH^- --->2 H_2O
Init amounts	10.0 mmol	
Add, mmol		10.005
Change	-10.0	-10.0
After rxn	~0	0.005

In this case, we can find the pOH by calculating the $[OH^-]$

$[OH^-] = 0.005/(20.0 + 20.01) = 1.25 \times 10^{-4}$ M

$pOH = -\log(1.25 \times 10^{-4}) = 3.9$

pH = 14.0 - pOH = 10.1

d. 20.10 mL added. Again we will have excess base. The amount of base is:

mmol OH^- = 0.5 x 20.10 = 10.05. After the reaction, 0.05 mmol of base will be left and $[OH^-] = 0.05/(20.0 + 20.1) = 1.25 \times 10^{-3}$ M

$pOH = -\log(1.25 \times 10^{-3}) = 2.9$

pH = 14.0 - 2.9 = 11.1

15. Titration curves

There are qualitative differences in the titration curve of a strong acid-strong base and weak acid-strong base.

The main qualitative differences in the two titrations are

1. A weak acid/strong base titration has a higher initial pH because the weak acid is only partially ionized.
2. At the half-neutralization point (halfway between the starting point and the equivalence point) in the titration of (or addition of) a strong base with a weak acid the pH = pK_a.
3. The pH > 7 at the equivalence point in a weak acid/strong base titration because the anion of the weak acid hydrolyzes. In a strong acid/strong base titration the equivalence point will be at pH = 7.
4. In a weak acid/strong base titration, the steep portion of the curve spanning the equivalence point is limited to a much smaller pH range. Because of this, the choice of indicators is more limited.

Exercise (text) 15.21A. For the titration described in Example 15.21, determine the pH after the addition of the following volumes of 0.500 M NaOH

a. 12.50 mL

b. 20.10 mL

Solution:

a. From Example 15.21, the amount of CH_3COOH is 10 mmol. Calculating the amount of species after the reaction:

	CH_3COOH	+	OH^- <-->	H_2O +	CH_3COO^-
Init. amount	10.00 mmol				
Add, mmol			6.25		
Change	-6.25		-6.25		+6.25
After rxn	3.75		~0		6.25

The simplest approach is to use the Henderson-Hasselbach equation and the concentrations using the total volume after the additions.

$$pH = pK_a + \log\frac{[CH_3COO^-]}{[CH_3COOH]} = 4.74 + \log(0.192/.115) = 4.74 + \log(1.67)$$

$= 4.96$

b. Adding 20.10 mL of 0.500 M NaOH. The amount of NaOH will be 10.05 mmol. This means that OH^- will be in excess and the pH will be the result of the excess. After the reaction, 0.05 mmol of OH^- will remain. The total volume is 40.10 mL.

$[OH^-] = 0.05$ mmol/40.10 mL $= 1.25 \times 10^{-3}$ M

$pOH = -\log(1.25 \times 10^{-3}) = 2.9$

$pH = 14.0 - pOH = 14.0 - 2.9 = 11.1$

16. Lewis acids and bases

A Lewis acid is an electron pair acceptor. A Lewis base is an electron pair donor. In a Lewis acid-base reaction, new covalent bonds are formed.

Quiz A

1. A proton donor is
   a. a Bronsted-Lowry base
   b. a Bronsted-Lowry acid
   c. insoluble
   d. a polyprotic acid
   e. none of these
2. Which acid is stronger: HBr or HCl?
3. Is an aqueous solution of $NaNO_3$ acidic, basic or neutral?
4. For the reaction $CN^- + H_2O \leftrightarrow HCN + OH^-$ determine both conjugate acid-base pairs. Be sure to label which is the acid and which is the base.
5. pOH = 4.0 is equivalent to
   a. an acidic solution
   b. pH = 10.0
   c. a very strong basic solution
   d. $[H_3O^+] = 10^{-4}$
   e. none of these
6. According to Bronsted-Lowry theory, in the reaction of ammonia with aqueous solution, what is the acid and what is the base? Write the equation for the reaction.
7. Is an aqueous solution of HOCl acidic, neutral or basic?

8. Acetic acid has a pK_a of 4.74 and phenol has a pK_a of 10.00. Which is the stronger acid?
9. A solution of 10^{-8} M HCl has a pH less than 7.0. Since pH = -log [H^+], why isn't the pH = 8?
10. What is a pH indicator?
11. What is the pH at the equivalence point when 20.00 mL of 0.500 M CH_3COOH is titrated with 0.500 M NaOH?

Quiz B

1. What is the pH of 0.020 M HCl solution?
2. Define a Bronsted-Lowry base.
3. Will an aqueous solution of Na_2CO_3 be acidic, basic or neutral?
4. Will the pH of 10^{-8} M HCl be greater than, less than or equal to 7.0?
5. The pH of a solution is 5.0. What is the pOH?
6. What is the pH of 0.020 M NaOH solution?
7. Define the equivalence point of a titration.
8. Would you expect a strong acid to have a larger or smaller value of pK_a than a weak acid?
9. Would you expect an aqueous solution of KCl to be acidic, neutral, or basic?
10. For the following reaction, identify the Bronsted-Lowry acids and bases and their conjugates:

    $NH_3 + H_2PO_4^- \longleftrightarrow NH_4^+ + HPO_4^{2-}$

Quiz C

True or False

1. Pure water is a nonelectrolyte and does not ionize significantly.
2. A Lewis acid is a proton donor.
3. The pH of a solution is indicated by the negative logarithm of the hydrogen ion concentration.
4. In water solution at 25 °C, pH + pOH = 12.
5. The stronger the acid, the more completely it ionizes in water.
6. A strong base is weakly ionized in water.
7. The conjugate base of a strong acid is a weak base.
8. A buffer solution will not change pH significantly when a small amount of an acid or base is added.
9. A molecular species can be a Lewis base and not a Bronsted-Lowry base.

Self Test

Match the following:

1. acid-base indicator
2. proton donor
3. Lewis acid
4. amphiprotic
5. Lewis base
6. proton acceptor
7. buffer solution
8. conjugate acid

a. a Bronsted-Lowry base
b. a Bronsted-Lowry acid
c. formed when a Bronsted-Lowry base accepts a proton
d. a substance added to a reaction mixture in a titration that changes color at or near the equivalence point
e. a solution containing a weak acid and its conjugate base or a solution containing a weak base and its conjugate acid
f. an electron pair acceptor
g. an electron pair donor
h. can ionize either as a Bronsted-Lowry acid or base, depending on the acid base properties of other species in solution

Problems/Short Answer

9. Butyric acid has a $K_a = 1.5 \times 10^{-5}$. What is the pH of an 0.15 M solution of butyric acid?
10. What is the $[OH^-]$ for a saturated solution of $Mg(OH)_2$ whose pH is 10.2?
11. Calculate the concentrations of all species present and the pH of a solution of 0.05 M hydroxylamine, NH_2OH, whose $K_b = 9.1 \times 10^{-9}$.
12. The pH of a solution is 5.6. What is the $[OH^-]$ concentration?
13. Which of the following acids is stronger, H_3PO_4 or $H_2PO_4^-$?
14. Which is a stronger acid, HF or HCl?
15. What is the $[H^+]$ concentration in a solution that is 0.01 M in NaOH?

CHAPTER 16

More Equilibria in Aqueous Solutions: Slightly Soluble Salts and Complex Ions

This chapter considers equilibria between slightly soluble salts and their ions in solution, and equilibria that involve complex ions. Many of the principles applied earlier will be used in this chapter to determine solubility and other behavior.

Chapter Objectives:

1. Define solubility product constant and be able to calculate concentrations from K_{sp}.
2. Describe why K_{sp} calculations are more subject to error than other equilibrium calculations.
3. Know how the common ion effect changes the solubility of ions and be able to calculate the concentration of ions when more than one source of ion occurs.
4. Know how to use the ion product quotient, Q_{ip}, to determine the direction of a precipitation reaction.
5. Determine whether precipitation is complete using the K_{sp} value.
6. Understand why and how pH will affect some precipitation reactions.
7. Define a complex ion and ligand and calculate equilibria concentrations when complex ions are present.
8. Explain the behavior of some complex ions to form acids in solution.
9. Write equations to describe the amphoteric nature of some metal hydroxides.
10. Describe the qualitative inorganic analysis scheme for common cations.

Chapter Summary

1. Solubility product constant, K_{sp}

The solubility product constant is the equilibrium constant for a solid in equilibrium with its ions in a saturated solution. We can calculate concentrations and the value of K_{sp} in the same way that we did for other equilibria. Remember that pure solids do not enter into the equilibrium expression so the amount of solid doesn't matter so long as some solid is present.

Example 16.1. Write a K_{sp} expression for equilibrium in a saturated solution of each of the following slightly soluble solutes: **a.** magnesium hydroxide **b.** copper (II) arsenate, $Cu_3(AsO_4)_2$.

Solution: The K_{sp} expression does not involve solids so it is the product of the ions, taking into account the stoichiometry of the reaction.

a. Write the ionization reaction:

$Mg(OH)_2(s) <--> Mg^{2+} + 2\ OH^-$

$K_{sp} = [Mg^{2+}][OH^-]^2$

b. Write the reaction:

$Cu_3(AsO_4)_2$ <--> $3\ Cu^{2+} + 2\ AsO_4^{3-}$

$K_{sp} = [Cu^{2+}]^3[AsO_4^{3-}]^2$

We knew from the name of the compound, Cu(II), that copper had a 2+ charge. Also, in both cases we need to know the stoichiometry of the reaction to write the correct K_{sp}.

2. Errors and K_{sp} calculations

K_{sp} calculations are more subject to error because interionic attractions can make the "effective" concentration (the activity) of an ion different from its stoichiometric concentration. This difference between activity and stoichiometric concentration is likely to be greatest for solutes that are moderately to highly soluble.

Example 16.2. The solubility of lead(II) fluoride listed in a handbook is 0.064 g PbF_2/100 mL H_2O at 20 °C. Calculate K_{sp} for PbF_2 at 20 °C.

Solution: To calculate K_{sp}, we need the concentration of ions in solution. We can calculate the concentration of PbF_2 and then use the ionization equation to get the concentration of ions.

$$\text{Mol } PbF_2 = \frac{0.064\,g}{100\,mL} \times \frac{1000\,mL}{1L} \times \frac{1 mol\ PbF_2}{245.2\,g} = 2.61 \times 10^{-3} \text{ mol}$$

PbF_2 <--> $Pb^{+2} + 2\ F^-$

So for each mol PbF_2, we have 1 mol Pb^{2+} and 2 mol F^-

$K_{sp} = [Pb^{+2}][F^-]^2 = (2.61 \times 10^{-3})(2 \times 2.61 \times 10^{-3})^2 = 7.1 \times 10^{-8}$

Exercise (text) 16.2A. In Example 15.4 we determined the molar solubility of magnesium hydroxide from the measured pH of its saturated solution. Use data from that example to determine the K_{sp} for $Mg(OH)_2$.

Solution: To calculate K_{sp} we need the concentration of OH^- and Mg^{2+}. From Example 17.3, we calculated the $[OH^-] = 3.3 \times 10^{-4}$ M. This will be two times $[Mg^{2+}]$ since the only source of OH^- is the ionization and the stoichiometry gives 2 OH^- for each Mg^{2+}. The $K_{sp} = [OH^-]^2[Mg^{2+}] = (3.3 \times 10^{-4})^2(1.7 \times 10^{-4}) = 1.9 \times 10^{-11}$.

Exercise (text) 16.3A. Calculate the molar solubility of silver arsenate, given that

$Ag_3AsO_4(s)$ <--> $3\ Ag^+(aq) + AsO_4^{3-}(aq)$ $\qquad K_{sp} = 1.0 \times 10^{-22}$

Solution: We can calculate the molar concentration of either ion from the K_{sp} value and, using the reaction stoichiometry, calculate the moles of silver arsenate that dissolved. From this we can express the solubility in g/100 mL if we need to. In this case, we are asked for molar solubility so the last conversion is unnecessary.

$$K_{sp} = [Ag^+]^3[AsO_4^{3-}]$$

Using s as the molar concentration of AsO_4^{3-} we can write

$$K_{sp} = (3s)^3s = 27s^4 = 1.0 \times 10^{-22}$$

$$s = 1.4 \times 10^{-6} = \text{moles of } Ag_3AsO_4/L$$

Exercise (text) 16.4A. Refer to Table 16.1 and arrange the following solutes in order of increasing molar solubility: MgF_2, CaF_2, $PbCl_2$, PbI_2.

Solution: In this case it is easy to estimate the molar solubility because each of these compounds will yield the same type of ionization, and solving for the solubility will involve taking the cube root of 1/4 K_{sp} in all cases.

$$MX_2 \text{ <--> } M^{2+} + 2X^-$$

$$K_{sp} = [M^{2+}][X^-]^2 = (s)(2s)^2 = 4s^3$$

We can compare the size of K_{sp} directly to determine the solubility. The solubility will be $PbCl_2 > MgF_2 > PbI_2 > CaF_2$.

3. Common ion effect

The common ion effect occurs when the solubility of a slightly soluble ionic compound is lowered in the presence of a second solute that furnished a common ion. This can be looked at as another application of Le Chatelier's principle. Effectively the concentration of one of the products has been increased by the common ion. The system will react to partially offset this change and more solid will form (less solubility). The equilibrium shifts to the left.

Exercise (text) 16.5A. Calculate the molar solubility of Ag_2SO_4 in 1.00 M $AgNO_3$(aq).

Solution: We have a common ion, Ag^+, from two sources. We can use the K_{sp} expression and the solubility equation to solve for the concentration of SO_4^{2-} and use this to determine the solubility.

	Ag_2SO_4(s)	<-->	$2Ag^+$ +	SO_4^{2-}
Initial conc. ($AgNO_3$)	1.00 M			
Dissolves			2s	s
Equilibrium			1.00 + 2s	s

$$K_{sp} = [Ag^+]^2[SO_4^{2-}] = (1.00 + 2s)^2(s) = 1.4 \times 10^{-5}$$

We can simplify by assuming that 2s is small compared to 1.00 M. Then $s = 1.4 \times 10^{-5}$ M = molar solubility.

4. Determining whether precipitation will occur: reaction quotient

The reaction quotient in the form of Q_{ip} is based on the initial concentrations. We can evaluate Q_{ip} in the following way: If $Q_{ip} > K_{sp}$, precipitation should occur. If $Q_{ip} < K_{sp}$ precipitation cannot occur, and if $Q_{ip} = K_{sp}$, the solution is just saturated.

Example 16.3. If 1.00 g $Pb(NO_3)_2$ and 1.00 g MgI_2 are both added to 1.50 L of H_2O, will a precipitate form?

$PbI_2(s) \leftrightarrow Pb^{2+}(aq) + 2\,I^-(aq) \qquad K_{sp} = 7.1 \times 10^{-9}$

Solution: Using the K_{sp} expression and the ion product quotient, Q_{ip}, we can determine whether a precipitate will form. We must first convert the amount to molar concentrations.

1.00 g $Pb(NO_3)_2 \times$ 1mol/331.2 g = 3.02×10^{-3} mol and the volume is 1.50 L. The molar concentration is 3.02×10^{-3} mol/1.5 L = 2.01×10^{-3} M.

1.00 g $MgI_2 \times$ 1 mol/278.1 g = 3.60×10^{-3} mol and 3.60×10^{-3} mol/1.50 L = 2.40×10^{-3} M

$K_{sp} = [Pb^{2+}][I^-]^2 = 7.1 \times 10^{-9}$

$Q_{ip} = (2.01 \times 10^{-3})(2 \times 2.40 \times 10^{-3})^2 = 4.63 \times 10^{-8}$

(Note the factor of 2 in the I^- concentration because each mole of MgI_2 produces 2 moles of I^-.)

Q_{ip} is larger than K_{sp}; therefore the reverse reaction will occur and a precipitate should form.

Example 16.4. Exactly 100 mL of 0.020 M KI is mixed with 175 mL of 0.0025 M $Pb(NO_3)_2$. Will a precipitate of $PbI_2(s)$ form?

Solution: K_{sp} for the reaction is given in the previous example,

$K_{sp} = 7.1 \times 10^{-9}$

We need to find the molar concentrations in the new solution and then proceed as in the previous example.

0.020 mol/L KI x 0.100 L = 0.0020 mol

and the new volume is 275 mL = 0.275 L.

[KI] = 0.0020 mol/0.275 L = 0.0073 M

This will also be the $[I^-]$.

0.0025 mol/L $Pb(NO_3)_2 \times$ 0.175 L = 4.4×10^{-4} mol

4.4×10^{-4} mol/0.275 L = 0.0016 M = $[Pb^{2+}]$

$Q_{ip} = [Pb^{2+}][I^-]^2 = (1.6 \times 10^{-3})(7.3 \times 10^{-3})^2 = 8.5 \times 10^{-8}$

$Q_{ip} > K_{sp}$; therefore, a precipitate should form.

5. Determining if precipitation is complete using the K_{sp} value.

A slightly soluble compound can never by completely precipitated; however, if $\leq$ 0.1% is left in solution we consider precipitation to be complete.

Exercise (text) 16.9A. To a solution with $[Ca^{2+}]$ = 0.0050 M, we add sufficient solid ammonium oxalate to make the solution $[C_2O_4^{2-}]$ = 0.0100 M. Will precipitation of Ca^{2+} as $CaC_2O_4(s)$ be complete?

$CaC_2O_4(s) \longleftrightarrow Ca^{2+}(aq) + C_2O_4^{2-}(aq) \qquad K_{sp} = 2.7 \times 10^{-9}$

Solution: Assuming initially that all the calcium precipitates, we can then calculate the amount that redissolves and determine how much Ca^{2+} is in solution. There will be an excess of $C_2O_4^{2-}$.

$CaC_2O_4(s) \longleftrightarrow$	$Ca^{2+}(aq)$	+	$C_2O_4^{2-}(aq)$
Initial conc.	0.0050 M - 0.0050 M		0.0100 M - 0.0050 M
Change	s		s
Equilibrium	s		(0.0050 M + s)

Assuming that the s is << 0.0050 and (0.0050 + s) ~ 0.0050

$K_{sp} = [Ca^{2+}][C_2O_4^{2-}] = s(0.0050) = 2.7 \times 10^{-9}$

$s = 5.4 \times 10^{-7}$ M

The percentage of Ca^{2+} remaining in solution is:

$(5.4 \times 10^{-7} \text{ M})/(5.0 \times 10^{-3} \text{ M}) \times 100\ \% = 0.011\%$

which is less than 0.1% and therefore precipitation is assumed to be complete.

6. Effect of pH on Solubility

If an acid-base reaction is involved as the solute dissolves, the solubility will be affected by pH.

Example 16.5. You slowly add NaOH(aq) to an aqueous solution of 0.10 M $FeSO_4(aq)$. At what pH will $Fe(OH)_2(s)$ begin to precipitate?

$Fe(OH)_2(s) \longleftrightarrow Fe^{2+}(aq) + 2\ OH^-(aq) \qquad K_{sp} = 8.0 \times 10^{-16}$

Solution: If we assume that NaOH is concentrated so that no change in concentration results from dilution of the solution, we can assume that the $FeSO_4$ concentration doesn't change. Using the concentration and K_{sp}, we can calculate the equilibrium concentration of OH^- and from that the pH.

$K_{sp} = [Fe^{2+}][OH^-]^2 = (0.10 \text{ M})(s)^2 = 8.0 \times 10^{-16}$

$s = 8.9 \times 10^{-8}$ M

$pOH = -\log(8.9 \times 10^{-8}) = 7.05$

$pH = 14.00 - pOH = 6.95$

At about neutral pH, precipitation will begin to occur.

Exercise (text) 16.12A. Describe the dissolution of $Mg(OH)_2(s)$ in $NH_4Cl(aq)$ through a net acid-base reaction in which NH_4^+ is the acid.

Solution: The two equilibrium equations are:

$Mg(OH)_2 <--> Mg^{2+} + 2\ OH^-$

and

$NH_4^+ + H_2O <--> NH_3 + H_3O^+$

The OH^- and H_3O^+ will react to form H_2O.

The net reaction will be

$NH_4^+ + OH^- <--> NH_3 + H_2O$

or

$Mg(OH)_2(s) + 2\ NH_4^+ <--> 2\ NH_3 + 2\ H_2O + Mg^{2+}$

7. Complex ions and ligands

A complex ion is a polyatomic cation or anion that consists of a central cation to which are bonded other groups (ligands).

To calculate the concentration of species, we use K_f, the formation constant, which is the equilibrium constant for the reversible reaction by which a complex ion is formed.

Exercise (text) 16.13A. Calculate the concentration of free silver ion, $[Ag^+]$, in an aqueous solution that is 0.10 M $AgNO_3$ and 1.0 M $Na_2S_2O_3$.

$Ag^+(aq) + 2\ S_2O_3^{2-}(aq) <--> [Ag(S_2O_3)_2]^{3-}(aq) \quad K_f = 1.7 \times 10^{13}$

Solution: We can solve for the Ag^+ concentration by using the K_f value and assuming that the reaction goes almost to completion.

	$Ag^+(aq)$ +	$2\ S_2O_3^{2-}(aq)$ <-->	$[Ag(S_2O_3)_2]^{3-}(aq)$
Initial conc.	0.1	1.0	
Change	-0.1	-0.2	+0.10
Final	~0	0.8	0.10

Although we have said that the concentration of Ag^+ is ~0, there is a small amount that we can determine from the K_f expression.

$$K_f = \frac{[NH_4^+][OH^-]}{[NH_3]} = \frac{x^2}{0.001} = \frac{0.1}{[Ag^+](0.8)^2} = 1.7 \times 10^{13}$$

$$[Ag^+] = \frac{0.1}{(1.7 \times 10^{13})(0.8)^2} = 9 \times 10^{-15}\ M$$

Exercise (text) 16.14A. If 1.00 g KI is added to 1.00 L of the solution described in Exercise 16.13A, should any AgI(s) precipitate from the solution?

$AgI(s) <--> Ag^+(aq) + I^-(aq)$ $K_{sp} = 8.5 \times 10^{-17}$

Solution: We can solve this by comparing the Q_{ip} value with K_{sp}. First we need to find the molar concentration of KI.

[KI] = 1.00g/L x 1 mol/166.0 g = 0.00602 M

$Q_{ip} = [Ag^+][I^-] = (9.2 \times 10^{-15})(6.02 \times 10^{-3}) = 5.5 \times 10^{-17}$

$Q_{ip} < K_{sp}$, so no precipitate will form.

8. Complex ions in acid-base reactions.

Some aqueous complex ions exhibit acidic properties. The small highly charged central atom withdraws electrons from the OH bond of water, allowing the release of hydrogen ions.

$[Al(H_2O)_6]^{3+} + H_2O <--> [Al(OH)(H_2O)_5]^{2+} + H_3O^+$

9. The amphoteric nature of some metal hydroxides.

An amphoteric hydroxide is one that can react with either an acid or a base. Examples are $Al(OH)_3$, $Zn(OH)_2$ and $Cr(OH)_3$.

The reaction with acid can be written:

$Cr(OH)_3(s) + 3\ H_3O^+ <--> [Cr(H_2O)_6]^{3+}$

The reaction with base is

$Cr(OH)_3(s) + OH^-(aq) <--> [Cr(OH)_4]^-(aq)$

10. Qualitative inorganic analysis scheme for common cations.

Fig. 16.13 of the book gives this scheme.

Skills Test

The polydentate ligand, EDTA, whose structure is

$[(OOCCH_2)_2NCH_2CH_2N(CH_2COO)_2]^{4-}$

can be used to chelate lead from the blood. The calcium EDTA complex is given by injection. The more stable Pb-EDTA complex is eliminated in the urine.

A child eats 10.0 g of paint containing 5.0% Pb by mass. How many grams of the calcium salt of EDTA should be given to chelate all of the lead?

Solution

10.0 g lead x 0.05 x 1 mole/207.2g = 2.41×10^{-3} mole Pb

Therefore, 2.41×10^{-3} mole of the calcium salt of EDTA should be given.

2.41×10^{-3} mole Pb x 1 mol Ca_2EDTA/mol Pb x 368.38 g/mol = 0.888 g Ca_2EDTA

Quiz A

1. Write the solubility product expression for $Ca_3(PO_4)_2$.
2. A saturated solution of a compound, MX, has a concentration of 9.0×10^{-6} M. What is the value of K_{sp}?
3. Which would be most effective in dissolving Ag_2SO_4 ?
   a. add NaOH
   b. add HCl
   c. add AgCl
   d. add H_2SO_4
   e. none of these
4. The effect of adding $FeCl_3$ to a solution of $Fe(OH)_3$ will be:
   a. an increase the concentration of OH^-
   b. a decrease the concentration of OH^-.
   c. no effect on the concentration of OH^-
   d. cannot tell from the information given.
5. Precipitation is said to be complete when
   a. the concentration of a substance is 0.0.
   b. the concentration of a substance is 0.1% of its initial concentration.
   c. no change in concentration occurs.
   d. the substance cannot be precipitated.
   e. none of these.
6. If $Q_{ip} = K_{sp}$, the solution
   a. will form a precipitate.
   b. cannot form a precipitate.
   c. will not react .
   d. is just saturated.
   e. is supersaturated.
7. Write the K_{sp} expression for lead (II) fluoride.

8. The K_{sp} for lead (II) fluoride is 7.1×10^{-8}. What is the molar solubility of PbF_2 ?
9. Write the solubility equilibrium equation described by the K_{sp} expression:

   $K_{sp} = [Pb^{2+}]^3[AsO_4^{3-}]^2$.
10. Calculate the molar solubility of $Mg(OH)_2$ in 0.050 M $MgCl_2$. K_{sp} of $Mg(OH)_2 = 1.8 \times 10^{-11}$.

Quiz B

1. The solubility of Ag_2SO_4 is 0.47 g/100 mL. What is the value of K_{sp}?
2. Write the solubility product expression for Bi_2S_3.
3. When the ion product quotient, Q_{ip}, is less than K_{sp}:

   a. more solid dissolves

   b. the reaction cannot occur

   c. more solid forms

   d. none of these
4. In which solution will $PbSO_4$ be most soluble?

   a. 0.100 M NaCl

   b. 0.100 M HCl

   c. 0.100 M H_2SO_4

   d. cannot be determined

   e. equally soluble in all solutions
5. A saturated solution of MX_2 has a concentration of 1.5×10^{-6} M. What is the value of K_{sp}?
6. Write the solubility equilibrium equation described by $K_{sp} = [Cr^{3+}][F^-]^3$.
7. What is the molar solubility of PbI_2 in 0.10 M KI? K_{sp} for $PbI_2 = 7.1 \times 10^{-9}$.
8. Define complex ion.
9. If $Q_{ip} > K_{sp}$, what happens when the equilibrium is developed in the solution?
10. Define formation constant, K_f.

Quiz C

Fill in the blank.

1. The equilibrium constant for the reaction in which a solid salt dissolves is called ________________.
2. The ion product (Q) will be ________________ to the K_{sp} for a saturated solution.
3. The effect of adding a common ion to one already in equilibrium in solution is to ____________ the solubility of the salt.
4. The solubility product expression for CaF_2 is $K_{sp} =$ ________________________.

5. ____________________ are soluble in both acidic and basic solutions.
6. If the ion product is greater than K_{sp}, ____________________ occurs.
7. The ____________________ measures the stability of a complex ion.
8. Selective precipitation of ions is used in ____________________.
9. Solubility of an ionic compound ____________________ if the solution contains a substance that can bond to a metal cation.
10. Addition of HCl to a solution of unknown metal ions causes ____________________ ions to precipitate.

Self Test

1. What is the solubility of $PbCl_2$ in 0.10 M NaCl ($K_{sp} = 1.6 \times 10^{-5}$)?
2. What is the solubility of $Al(OH)_3$ in 0.10 M NaCl ($K_{sp} = 1.3 \times 10^{-33}$)?
3. How would you expect addition of HCl to affect the solubility of $Al(OH)_3$?
4. The molar solubility of $Mg(OH)_2$ is 1.12×10^{-4} M. What is the value of K_{sp}?
5. Would you expect the solubility of $Mg(OH)_2$ to be greater at pH 8.0 or pH 9.0 ? Why?
6. Calculate the ratio of $[Ag^+]/[Ag(CN)^{2-}]$ in a NaCN solution that is 0.02 M in $[CN^-]$. $K_f = 1.0 \times 10^{21}$.
7. Calculate the solubility of $Al(OH)_3$ at pH 8.0 ($K_{sp} = 1.3 \times 10^{-33}$).
8. Which is the more stable complex ion: $[Co(NH_3)_6]^{3+}$, $K_f = 4.5 \times 10^{33}$ or $[Cu(NH_3)_4]^{2+}$, $K_f = 1.1 \times 10^{13}$?
9. AgI is most soluble in which of the following ?

   a. 0.01 M NaI

   b. 0.1 M $AgNO_3$

   c. 0.2 M KI

   d. pure water
10. Will a precipitate form when 10.0 mL of a solution at pH 9.0 is mixed with 10.0 mL of a 1.0×10^{-4} M solution of $FeCl_3$ ($Fe(OH)_3$ $Ksp = 4 \times 10^{-38}$)?

CHAPTER 17

Thermodynamics: Spontaneity, Entropy, and Free Energy

Thermodynamics can help us determine yields, stability of substances, and optimum temperature and pressure for reactions. Thermodynamics can also describe the conversion of energy and work.

Chapter Objectives: You should be able to:

1. Define a spontaneous process.
2. Describe the thermodynamic property related to disorder of a system.
3. Know the relationship between entropy, heat and temperature.
4. State the conditions under which the entropy is zero.
5. State the second law of thermodynamics.
6. Write the equation for free energy change in a process at constant temperature and relate the change in free energy to the spontaneity of a reaction or process at constant temperature and pressure.
7. Know how to use standard free energies of formation to predict the spontaneity of reactions.
8. Calculate the equilibrium constant from the free energy of a reaction.
9. Relate the equilibrium constant, K_{eq}, at two temperatures using the van't Hoff equation.

Chapter Summary

1. Spontaneous process

A spontaneous process is one that can occur in a system left to itself; no action from outside the system is necessary. An example of a spontaneous change is the melting of ice at room temperature.

Example 17.1. Use your general knowledge to indicate whether each of the following processes is spontaneous or nonspontaneous. Comment on cases for which you cannot make a clear determination.

- **a.** The decay of a piece of lumber buried in soil
- **b.** The formation of sodium, Na(s) and chlorine, Cl_2, by vigorous stirring of an aqueous solution of sodium chloride, NaCl(aq)
- **c.** The formation of lime, CaO(s), and carbon dioxide, CO_2(g), from limestone, $CaCO_3$(s), at 600 °C

Solution: Our definition of a spontaneous process is one that occurs without further action. That is, when left to itself, the process will occur.

- **a.** The decay of a piece of lumber buried in soil is a spontaneous process because the wood eventually oxidizes. The decay of the lumber is enhanced by the action of bacteria.

b. Na^+ and Cl^- ions are stable in solution and stirring will not cause the reaction to reverse. The process is nonspontaneous. We can also say that dissolving NaCl in aqueous solution (the reverse reaction) is spontaneous.

c. Indeterminate. $CaCO_3$ will decompose at high temperature, but whether this will occur at 600 °C, we cannot say.

2. Entropy

The thermodynamic property related to the disorder of a system is the entropy. Entropy is a state function and depends only on the present conditions of the system, not how much work was done to achieve the state. To determine whether an overall process is spontaneous, we must determine the energy and the entropy change. A decrease in energy and an increase in entropy both favor a process. Often these factors work in opposite directions and we must determine which factor predominates.

Exercise (text) 17.2A. Predict whether each of the following leads to an increase or decrease in entropy. If a prediction is not possible, explain why.

a. $NH_3(g) + HCl(g) \rightarrow NH_4Cl(s)$

b. $2\ KClO_3(s) \rightarrow 2\ KCl(s) + 3\ O_2(g)$

c. $CO(g) + H_2O(g) \rightarrow CO_2(g) + H_2(g)$

Solution: To make predictions, we need to keep in mind that the entropy for a solid < liquid << gas. Therefore, reactions that yield an increase in the number of moles of gas will increase the entropy.

a. In this reaction 2 mol gas--> 1 mole solid. The entropy will decrease.

b. Three moles of gas are produced by this reaction. The entropy will increase.

c. While the number of moles of gas is the same on each side of the equation, the gases found on the right have higher symmetry than those found on the left and are more ordered. Thus, there has been a small, but measurable decrease in the entropy. The is calculated in Example (text) 17.3A below.

3. The relationship between entropy, heat and temperature

The change in entropy is directly proportional to the heat, q, and inversely proportional to the Kelvin temperature, T.

$$\Delta S \propto \frac{q}{T}$$

The relationship between q and ΔS seems reasonable since if a system absorbs a large quantity of heat, more disorder will occur than if a small quantity of heat is absorbed. The inverse relationship with temperature is less obvious. However, if a system is already highly disordered at high temperature, additional heat will create less disorder than if the same quantity of heat is absorbed at lower temperature.

4. Standard Molar Entropy

The third law of thermodynamics states that the entropy of a pure, perfect crystal is zero at a temperature of 0 K.

Exercise (text) 17.3A. Use data from Appendix C to calculate the standard molar entropy change at 25 °C for the following reaction.

$CO(g) + H_2O(g) \rightarrow CO_2(g) + H_2(g)$

Solution. To calculate the standard molar entropy change we use the formula

$$\Delta S^\circ = \sum \nu_p \times S^\circ_{prod} - \sum \nu_r \times S^\circ_{react}$$

$$= S^0_{CO_2(g)} + S^0_{H_2(g)} - S^0_{CO(g)} - S^0_{H_2O(g)}$$

$$= 213.6\ Jmol^{-1}K^{-1} + 130.6\ Jmol^{-1}K^{-1} - 197.6\ Jmol^{-1}K^{-1} - 188.7\ Jmol^{-1}K^{-1}$$

$$= -42.1\ Jmol^{-1}K^{-1}$$

5. The Second Law of Thermodynamics

The second law of thermodynamics states that all spontaneous processes increase the entropy of the universe.

6. Free Energy and Free Energy Change

$$\Delta G = \Delta H - T\Delta S$$

We can use ΔG to predict spontaneous changes at constant T and P.

If $\Delta G < 0$ (negative), a process is spontaneous.

If $\Delta G > 0$ (positive), a process is nonspontaneous.

If $\Delta G = 0$, a process is at equilibrium.

An examination of the equation relating free energy, entropy and enthalpy will allow us to estimate qualitatively whether or not a reaction is spontaneous.

Table 17.1 in the text book gives criteria for spontaneous change. We need to determine whether ΔH or $-T\Delta S$ will dominate in a reaction, or if both terms will favor the same direction of the reaction. For example, if ΔH is negative and ΔS positive, both terms will be negative and ΔG will be negative (spontaneous reaction) at all temperatures. If ΔH is positive and ΔS is negative, the reaction will not be spontaneous at any temperature. For other cases, one must consider the magnitude of the $T\Delta S$ term depending on the temperature.

Exercise (text) 17.4A. Predict which of the four cases in Table 17.1 are likely to apply to the following reactions.

a. $N_2(g) + 2\ F_2(g) \rightarrow N_2F_4(g)$ $\Delta H = -7.1$ kJ

b. $COCl_2(g) \text{-->} CO(g) + Cl_2(g)$ $\Delta H = +110.4$ kJ

Solution: To determine whether a change is spontaneous, we use the equation:

$\Delta G = \Delta H - T\Delta S$

a. ΔH is negative and, since we are producing fewer moles of gas, ΔS will also be negative. The reaction will be nonspontaneous toward high temperature.

b. ΔH is positive and ΔS will also be positive since the moles of gas are increasing. The reaction will be spontaneous toward high temperature.

7. Using standard free energy change to predict the spontaneity of a reaction

$$\Delta G = \sum \nu_p \Delta G_f^0(\text{products}) - \sum \nu_r \Delta G_f^0(\text{reactants})$$

where ν is the stoichiometric coefficient and the calculation using standard free energies of formation is the same type of calculation used for standard enthalpies in Chapter 5.

Example 17.2. Determine the standard free energy change at 25 °C for these reactions:

a. $2\ NO(g) + O_2(g) \text{-->} 2\ NO_2(g)$ $\Delta H^o = -114.1$ kJ $\Delta S^o = -146.2$ J K^{-1}
using the Gibbs equation.

b. $CS_2(l) + 2\ S_2Cl_2(g) \text{-->} CCl_4(l) + 6\ S(s)$
using standard free energies of formation.

Solution.

a. Using the Gibbs equation, $\Delta G^o = \Delta H^o - T\Delta S^o$. We must convert ΔS^o to kJ to substitute into the equation.

$\Delta G^o = \Delta H^o - T\Delta S^o = -114.1\text{ kJ} - (298\text{ K})(-0.1462\text{ kJ K}^{-1})$

$= -70.4$ kJ

b. Using standard free energies of formation,

$\Delta G^o = \Delta G_f^o[CCl_4(l)] + 6\ \Delta G_f^o[S(s)] - \Delta G_f^o[CS_2(l)] - 2\ \Delta G_f^o[S_2Cl_2(g)]$.

Substituting values from Appendix C into the equation,

$\Delta G^o = -65.27\text{ kJ} + 6(0)\text{ kJ} - 65.27\text{ kJ} - 2(-31.8\text{ kJ}) = -66.9$ kJ

8. Calculating the equilibrium constant from the free energy of a reaction.

$\Delta G^o = -RT \ln K_{eq}$

Exercise (text) 17.8B. Write an equation for the dissolution of magnesium hydroxide in an acidic solution, and then write the K_{eq} expression for this reaction.

Solution: The equation for dissolution is:

$Mg(OH)_2(s) + 2\ H^+(aq) \text{-->} Mg^{2+}(aq) + 2\ H_2O(l)$

$K_{eq} = \frac{[Mg^{2+}]}{[H^+]^2}$ No term will appear for solid or water; each is a pure phase with a = 1.

Exercise (text) 17.9A. Use data from Appendix C to determine the value of K_{eq} at 25 °C for the reaction: 2 $HgO(s) \longleftrightarrow 2\ Hg(l) + O_2(g)$

Solution. We can use the data from the appendix to calculate ΔH^o and ΔS^o. From these values we can then calculate ΔG^o and K_{eq}. We can also calculate ΔG^o from ΔG_f^o values. Using the latter approach,

$$\Delta G^o = \Delta G_f^o\ [O_2] + 2\Delta G_f^o[Hg] - 2\Delta G_f^o\ [HgO]$$
$$= 0\ kJ + 2 \times 0\ kJ - 2 \times (-58.56\ kJ) = 117.12\ kJ$$
$$\Delta G^o = -RT\ \ln K_{eq}$$
$$117.12\ kJ = -(8.3145\ J\ mol^{-1}\ K^{-1})\ (298.15\ K)\ \ln K_{eq}$$
$$\ln K_{eq} = 1.1712 \times 10^5\ J/(-8.3145\ J\ mol^{-1}\ K^{-1})\ (298.15\ K) = -47.245$$
$$K_{eq} = 3.03 \times 10^{-21}$$

9. The dependence of ΔG^0 and K_{eq} on Temperature

The van't Hoff equation relates the equilibrium constant at two temperatures:

$$\ln\frac{K_2}{K_1} = \frac{\Delta H^o}{R}\left(\frac{1}{T_1} - \frac{1}{T_2}\right)$$

The assumption in writing this equation is that ΔH^o and ΔS^o are independent of temperature.

Exercise (text) 17.10A. In Example 17.9, we determined the value of K_{eq} at 25 °C for the reaction 2 $NO_2(g) \longleftrightarrow N_2O_4(g)$. What is the value of K_{eq} for the same reaction at 65 °C?

Solution: We can determine ΔH^o from tabulated data and then apply the van't Hoff equation to determine the new K_{eq}.

$$\Delta H^o = \Delta H^o[N_2O_4(g)] - 2\ \Delta H^o[NO_2(g)]$$
$$= 9.16\ kJ - 2 \times 33.18\ kJ = -57.20\ kJ$$

Using the van't Hoff equation:

$$\ln\frac{K_2}{K_1} = \frac{\Delta H^o}{R}\left(\frac{1}{T_1} - \frac{1}{T_2}\right) = \frac{-5.720 \times 10^4\ J}{8.3145\ J\ mol^{-1}\ K^{-1}}\left(\frac{1}{298} - \frac{1}{338}\right) = -2.73$$

$$\frac{K_2}{K_1} = e^{-2.73} = 0.065$$

$$K_2 = 0.065 \times K_1 = 0.065 \times 6.9 = 0.45$$

Tabulated values of ΔG_f^o, ΔH_f^o , and S^o are generally for 25 °C. To use these data to calculate values for K_{eq} at other temperatures requires the assumption that ΔH_f^o and S^o are independent of temperature. We can then use the van't Hoff equation to relate the equilibrium constant and temperature.

Skills Test Problem

Calculate the pH of pure water at 37 °C.

Solution

First calculate the ionization constant of water at 37 °C.

$$\ln\frac{K_2}{K_1} = -\frac{\Delta H^0}{R}\left(\frac{1}{T_2} - \frac{1}{T_1}\right)$$

The ionization for water at 25 °C is 1.00×10^{-14} and $\Delta H^0 = 55.84$ kJ

$K_1 = 10^{-14}$, $T_1 = 298$ and $T_2 = 310$

$$\ln\frac{K_2}{10^{-14}} = -\frac{55.840}{8.314}\left(\frac{1}{310} - \frac{1}{298}\right)$$

$K_2 = 2.4 \times 10^{-14}$ which is the ionization constant of water at 37 °C.

$[H^+][OH^-] = 2.4 \times 10^{-14}$

$[H^+] = [OH^-]$

$= (2.4 \times 10^{-14})^{1/2} = 1.55 \times 10^{-7}$

$pH = -\log[H^+] = 6.81$

Quiz A

1. Predict whether ΔS is positive, negative or impossible to predict.

   $NH_3(g) + HCl(g) \rightarrow NH_4Cl(s)$

2. The free energy change for a reaction is -14.5 kJ. The reaction is
   a. exothermic
   b. endothermic
   c. nonspontaneous
   d. spontaneous
   e. at equilibrium

3. The energy available for work from a reaction is

a. entropy
b. enthalpy
c. free energy
d. always increasing
e. all of these

4. A given exothermic reaction increases the order of the system. Under what conditions is the reaction spontaneous?
   a. low temperature
   b. high temperature
   c. all temperatures
   d. never spontaneous
5. For the reaction 2 $NO(g) + Cl_2(g) \rightarrow 2 NOCl(g)$ at 25 °C, $\Delta H^o = -37.78$ kJ/mol and $\Delta S^o = -117.03\ J\ mol^{-1}\ K^{-1}$. What is the value of ΔG^o for the reaction?
6. Define free energy.
7. Under what conditions is the entropy of a system zero?
8. For the reaction $CCl_4(l) \rightarrow CCl_4(g)$, predict whether ΔS is positive, negative, or a prediction cannot be made.
9. What must be the temperature if a reaction has $\Delta G = -615$ kJ, $\Delta H = -852$ kJ and $\Delta S = -145\ JK^{-1}$?
10. What is meant by a reaction being nonspontaneous?

Quiz B

1. For the reaction $SO_2Cl_2(l) \rightarrow SO_2(g) + Cl_2(g)$, predict whether ΔS is positive, negative, or a prediction cannot be made.
2. For an exothermic reaction,
   a. $\Delta H > 0$
   b. $\Delta H < 0$
   c. the reaction is spontaneous
   d. the reaction occurs rapidly
   e. the reaction is nonspontaneous
3. A reaction can be written that is both endothermic and increases the order of the system. The reaction would be:
   a. spontaneous at low temperature
   b. spontaneous at high temperature
   c. spontaneous at all temperatures
   d. nonspontaneous at all temperatures
   e. cannot predict the spontaneity of the reaction
4. Write the equilibrium constant expression for the reaction

$H_2(g) + Br_2(g)$ <--> 2 HBr(g).

5. ΔG^o for a reaction at 25 °C is -27.0 kJ, what is K_{eq} for the reaction?
6. What is meant by a spontaneous change?
7. For the reaction 2 NO(g) + $O_2(g)$ <--> 2 $NO_2(g)$, the greatest amount of $NO_2(g)$ is produced at low temperatures. What can be said about the enthalpy of the reaction?
8. Define entropy.
9. For a reaction at 200 K, $\Delta G = 250$ kJ and $\Delta H = 450$ kJ. What is the value of ΔS?
10. For a system at equilibrium, ΔG is
    a. positive
    b. negative
    c. zero
    d. equal to K_{eq}
    e. none of these

Quiz C

True or False

1. All spontaneous reactions are exothermic.
2. A spontaneous reaction moves a system towards equilibrium.
3. If a mole of O_2 molecules breaks down to form O atoms, there is a simultaneous increase in entropy.
4. The entropy of a particular state is related to the number of ways that state can be achieved.
5. Molecular systems tend to spontaneously become more ordered.
6. In any process, spontaneous or nonspontaneous, the total energy of the systems and its surroundings is constant.
7. A spontaneous reaction occurs rapidly.
8. Entropy is associated with molecular motion and therefore the entropy increases as the temperature increases.
9. Enthalpy alone accounts for the direction of a spontaneous change.

Self Test

Matching

1. entropy
2. third law of thermodynamics
3. second law of thermodynamics

a. the science dealing with the relationship between heat and work and with transformation of energy from one form to another

b. a process that occurs in a system left to itself.

c. measures the degree of randomness or

disorder in a system

4. pontaneous process

d. all natural or spontaneous processes are accompanied by an increase in entropy of the universe.

5. free energy

e. the entropy of a pure perfect crystal at 0 K is zero.

6. thermodynamics

f. a thermodynamic function used for establishing criteria for equilibrium and for spontaneous change

Problems/Short Answer

7. For the precipitation reaction

   $Ag^{+}(aq) + Cl^{-}(aq) \rightarrow AgCl(s)$ $\Delta H = -65$ kJ

   Since the entropy decreases, why is the reaction spontaneous?

8. For the given reaction,

   $Ca(OH)_2(s) \rightarrow CaO(s) + H_2O(g)$

   Given that $\Delta H[Ca(OH)_2] = -986.2$ kJ/mol and $S^0[Ca(OH)_2] = 83.4$ J/K mol and data in appendix C, determine at what temperature the reaction will be spontaneous.

9. What is the equilibrium constant for the reaction in problem 8 at 500 K?

10. Which of the following are state functions:

    a. temperature
    b. work
    c. density
    d. enthalpy
    e. heat

CHAPTER 18

Electrochemistry

This chapter deals with how spontaneous chemical reactions can produce electricity and how electricity can be used to produce nonspontaneous reaction.

Chapter Objectives:

1. Write and balance half-reactions for electrochemical reactions.
2. Describe a simple electrochemical cell and be able to label the electrodes, the oxidation and reduction half-cells. Be able to indicate the direction of electron flow in a cell.
3. Explain how a standard hydrogen electrode (SHE) is used to determine electrode potential.
4. Calculate cell potentials from electrode potentials.
5. Relate E^{o}_{cell} and the free energy for an electrochemical cell. Interpret the value of E_{cell} to determine the direction of a reaction and whether the cell is at equilibrium.
6. Determine whether a metal will displace other metal ions from solution.
7. Know how to determine K_{eq} from $E^{o}_{cell.}$
8. Apply the Nernst equation to nonstandard conditions to calculate $E_{cell.}$
9. Describe a concentration cell.
10. Define a battery and describe three types of cells or batteries.
11. Describe the process of corrosion.
12. Define electrolysis.
13. Determine the quantity of reactant consumed or product formed during electrolysis.

Chapter Summary

1. Half-Reactions

A galvanic cell is an electrochemical cell that produces electricity from an oxidation-reduction reaction. The two half-reactions are an oxidation reaction and a reduction reaction. The two half-reactions are combined into an over all oxidation-reduction reaction. In balancing these reactions we must remember that the same number of electrons involved in oxidation is involved in the reduction reaction.

The total electric charge that flows is expressed in coulombs, the rate of charge flow or electric current is expressed in amperes (coulombs per second).

Example 18.1. Balance the following equation using the half-reaction method.

$Ag^{+}(aq) + Fe(s) \rightarrow Fe^{3+}(aq) + Ag(s)$

Solution: Using the half-reaction method we write an oxidation and a reduction reaction. The number of electrons is balanced so that the same number of electrons is lost as is gained.

The half-reactions are:

$Fe(s) \rightarrow Fe^{3+}(aq) + 3e^{-}$ oxidation

$Ag^{+}(aq) + e^{-} \rightarrow Ag(s)$ reduction

To balance the half-reactions, we need to multiply the reduction reaction by 3 so that the electrons balance. The balanced reaction will be

$3Ag^{+}(aq) + Fe(s) \rightarrow Fe^{3+}(aq) + 3Ag(s)$

Example 18.2. Use the half-reaction method to balance the following equation:

$Cr_2O_7^{2-}(aq) + Cl^{-}(aq) \rightarrow Cr^{3+}(aq) + Cl_2(g)$ (acidic solution)

First, the two half reactions are

$Cl^{-}(aq) \rightarrow Cl_2(g)$

$Cr_2O_7^{2-}(aq) \rightarrow Cr^{3+}(aq)$

Second, we balance each half-reaction. Cl_2 requires two Cl

$2Cl^{-}(aq) \rightarrow Cl_2(g)$

and the two minus charges require two electrons on the right-hand side of the equation.

$2Cl^{-}(aq) \rightarrow Cl_2(g) + 2e^{-}$

For the reduction reaction, two Cr^{3+} are required to balance the Cr atoms and seven oxygen atoms are required. The oxygen can be accounted for by adding water to the right side of the equation

$Cr_2O_7^{2-}(aq) \rightarrow 2Cr^{3+}(aq) + 7H_2O(l)$

Because the reaction occurs under acidic conditions, we can add H^{+} to the left side of the equation to balance the H atoms.

$14H^{+} + Cr_2O_7^{2-}(aq) \rightarrow 2Cr^{3+}(aq) + 7H_2O(l)$

Charge is balanced by adding electrons to the left side.

$6e^{-} + 14H^{+} + Cr_2O_7^{2-}(aq) \rightarrow 2Cr^{3+}(aq) + 7H_2O(l)$

To equalize the number of electrons transferred, we multiply the first half-reaction by three. We now add the two half-reactions together to obtain the overall balanced equation.

$$6Cl^- (aq) \rightarrow 3Cl_2 (g) + 6e^-$$

$$\underline{6e^- + 14H^+ + Cr_2O_7^{2-} (aq) \rightarrow 2Cr^{3+} (aq) + 7H_2O (l)}$$

$$14H^+ + Cr_2O_7^{2-} (aq) + 6Cl^- \rightarrow 2Cr^{3+} (aq) + 3Cl_2 (g) + 7H_2O (l)$$

There are equal numbers of atoms on each side of the equation and the charge is the same on each side of the equation so the equation is balanced.

2. Cell Diagrams

1. Electrochemical cells are often represented by cell diagrams. The anode is at the left side of the diagram and the cathode is at the right side. A single vertical line represents the boundary between phases and a double line represents a salt bridge separating two half-cells.

Example 18.3. Write the net ionic equation for the redox reaction that occurs in the voltaic cell:

$Al(s) \mid Al^{3+}(aq) \parallel Cu^{2+}(aq) \mid Cu(s)$

Solution: To write the net equation, first write the half-reactions and balance them for oxidation-reduction reactions as we learned earlier. The half-cell on the left represents the oxidation reaction:

$Al(s) \dashrightarrow Al^{3+} + 3\,e^-$

The reduction reaction is:

$Cu^{2+} + 2\,e^- \dashrightarrow Cu(s)$

To balance the electrons, multiply the first equation by 2 and the second by 3:

$2\,Al(s) \dashrightarrow 2\,Al^{3+}(aq) + 6\,e^-$

$3\,Cu^{2+}(aq) + 6\,e^- \dashrightarrow 3\,Cu(s)$

The net equation is the sum of the two equations:

$2\,Al(s) + 3\,Cu^{2+}(aq) \dashrightarrow 2\,Al^{3+}(aq) + 3\,Cu(s)$

Example 18.4. Suppose a voltaic cell is composed of an anode half-cell in which Zn(s) is oxidized to $Zn^{2+}(aq)$ and a cathode half-cell where $Cl_2(g)$ is reduced to $Cl^-(aq)$. Write an equation for the net cell reaction and a cell diagram for the voltaic cell.

Solution: Write the two half-reactions and determine the net reaction. From these reactions, write the cell diagram.

For the anode:

$Zn(s) \text{ --> } Zn^{2+}(aq) + 2\,e^-$

For the cathode:

$Cl_2(g) + 2\,e^- \text{ --> } 2\,Cl^-(aq)$

The electrons are balanced, so the net reaction is:

$Zn(s) + Cl_2(g) \text{ --> } Zn^{2+}(aq) + 2\,Cl^-(aq)$

The cell diagram is:

$Zn(s)\,|\,Zn^{2+}(aq)\,||\,Cl^-(aq)\,|\,Cl_2(g), Pt$

3. Standard hydrogen electrode (SHE)

The standard electrode potential is based on the tendency for reduction to occur at the specified electrode (25 ^{0}C). All solution species are present at unit activity (a = 1), which is about 1 M, and gases at 1 atm pressure. The SHE is used as a reference electrode in electrochemical cells. The electrode potential for the SHE is set at exactly 0.00 volts. If the sign of the standard electrode potential for a reaction such as Cu^{2+} to Cu(s) is positive, that reduction is easier, and more spontaneous, than reduction of $[H^+]$ to H_2 (g).

4. Calculating cell potentials from electrode potentials

To calculate standard cell potentials, E^o_{cell}, follow the procedure:

1. Write the reduction half-equation and its standard potential, E^o_{red} (Table 18.1).
2. Write the oxidation half-equation and its standard potential, E^o_{oxid}, which is the negative of E^o_{red}.
3. Combine the half-equations and add the half-cell potentials: $E^o_{cell} = E^o_{red} + E^o_{oxid}$.

The coefficients of the half-reactions do not affect values of the cell potential.

Example 18.5. Use the value of E^o_{cell} from Table 18.1 for the reduction of Co^{2+}(aq) to Co(s), together with the following information:

$2\,Ce^{4+}(1M) + Co(s) \text{ --> } 2\,Ce^{3+}(1M) + Co^{2+}(1M) \quad E^o_{cell} = 1.887\text{ V}$

to determine E^o for the reduction half-reaction:

$Ce^{4+}(1M) + e^- \text{ --> } Ce^{3+}(1M)$

Solution: First, determine the oxidation and reduction half-reactions. We are given the reduction reaction. The oxidation reaction is Co(s) --> Co^{+2}(aq). Use the relationship $E^o_{cell} = E^o_{oxid} + E^o_{red}$. For the oxidation reaction, $E^o_{oxid} = +0.277$ V.

$E^o_{red} = E^o_{cell} - E^o_{oxid} = 1.887\ V - 0.277\ V = 1.610\ V.$

5. E^o_{cell}, free energy, and the direction of change.

Given the relationship between E^o_{cell} and ΔG, this interpretation is straightforward. The sign of the free energy tells us the direction of the reaction.

$$\Delta G^o = - n \times F \times E^o_{cell}$$

where F is the Faraday constant and n is the number of electrons transferred. F = 96,485 $Cmol^{-1}$. But since 1 J = 1 V x 1C, F = 96.485 $kJmol^{-1}V^{-1}$, a more convenient expression for calculating free energy.

Example 18.6. Will the following reaction occur spontaneously in the forward direction?

$$Cu^{2+}(aq) + 2\ Fe^{2+}(aq) \dashrightarrow 2\ Fe^{3+}(aq) + Cu(s)$$

Solution: To determine whether the reaction is spontaneous as written, determine E^o_{cell}. Using Table 18.1 find E^o_{red} for the two reactions. The reduction reaction is:

$Cu^{2+}(aq) + 2\ e^- \dashrightarrow Cu(s)$ and $E^o_{red} = +0.337\ V$.

The oxidation half-reaction is:

$2\ Fe^{2+}(aq) \dashrightarrow 2\ Fe^{3+}(aq) + 2\ e^-$ and $E^o_{oxid} = -0.771\ V$.

$E^o_{cell} = E^o_{oxid} + E^o_{red} = -0.771\ V + 0.337\ V = -0.434\ V$

Since E^o_{cell} is negative and $\Delta G^o = -n \times F \times E^o_{cell}$, ΔG^o will be positive. The reaction will not be spontaneous in the direction written.

6. Determining whether a metal will displace other metal ions from solution

A metal will displace from a solution of its ions any metal lying below it in the table of standard electrode potentials. That is, any metal having larger value for E^o_{red} will be displaced by one having a smaller E^o_{red} value.

7. Determining K_{eq} from E^o_{cell}

Rearranging the two equations:

$$\Delta G = - RT \ln K_{eq} = -n \times F \times E^o_{cell},$$

allows us to write

$$E^o_{cell} = \frac{RT \ln K_{eq}}{nF} = \frac{0.0592}{n} \log K_{eq} \quad \text{at } 25\ ^oC$$

Use the cell half-reactions to determine the number of electrons transferred, n, in the equation

Example 18.7. Calculate the value of K_{eq} at 25 °C for the reaction

$Cu(s) + 2\ Fe^{3+}(1M) <--> Cu^{2+}\ (1M) + 2\ Fe^{2+}(1M)$

Solution: Calculate E^o_{cell}, and from that, the value of K_{eq}. Determine the half-cell reactions to evaluate the value of n. The above reaction is the reverse of the reaction in Exercise 20.4, so the sign of E^o_{cell} will be reversed.

$$E^o_{cell} = E^o_{oxid} + E^o_{red} = +0.434\ V$$

From the combined half-cell reactions, n = 2 moles of e^- are transferred in the reaction.

$$E^o_{cell} = (0.0592/n) \times \log K_{eq} = 0.434\ V$$

$$\log K_{eq} = 0.434/(0.0592/2) = 14.7$$

$$K_{eq} = 4.5 \times 10^{14}$$

8. Applying the Nernst equation to nonstandard conditions to calculate E_{cell}

The Nernst equation is

$E_{cell} = E^o_{cell} - \frac{0.0592}{n} \log Q$ where Q is the reaction quotient.

Exercise (text) 18.9. Use the Nernst equation to determine E_{cell} at 25 °C for the following voltaic cells:

a. $Zn(s)\ |\ Zn^{2+}(2.0M)\ ||\ Cu^{2+}(0.050\ M)\ |\ Cu(s)$
b. $Zn(s)\ |\ Zn^{2+}(0.050M)\ ||\ Cu^{2+}(2.0\ M)\ |\ Cu(s)$
c. $Cu(s)\ |\ Cu^{2+}(1.0\ M)||\ Cl^-(0.25\ M)\ |\ Cl_2(g, 0.50\ atm),Pt$

Solution: To determine E_{cell}, write the half-reactions. Determine the expression for Q and substitute into the Nernst equation. The value of n must also be determined.

Note that the left side of the diagram is the anode, where oxidation occurs. The equation is

$$Zn(s) + Cu^{2+} <--> Zn^{2+} + Cu(s)$$

Two moles of electrons are transferred in the reaction.

$$E^o_{cell} = E^o_{oxid} + E^o_{red} = +0.763\ V + 0.337\ V = 1.100\ V$$

From the Nernst equation:

$$E_{cell} = E^o_{cell} - (0.0592/n) \log Q$$

$$= 1.100\ V - (0.0296) \log \frac{[Zn^{2+}]}{[Cu^{2+}]}$$

a. $[Zn^{2+}] = 2.0$ M and $[Cu^{2+}] = 0.050$ M Substituting into the above equation

$$E_{cell} = 1.100\text{ V} - (0.0296)\log\frac{2.0\text{M}}{0.050\text{M}} = 1.100\text{ V} - 0.0296\log(40) = 1.053\text{ V}$$

Since we have decreased one of the reactants (Cu^{2+}) and increased one of the products (Zn^{2+}), we would expect the voltage to decrease.

b. Using the equation above with different concentrations:

$$E_{cell} = E^o_{cell} - (0.0592/n)\log Q$$

$$= 1.100\text{ V} - (0.0296)\log\frac{[Zn^{2+}]}{[Cu^{2+}]}$$

$$= 1.100\text{ V} - (0.0296)\log\frac{0.050\text{ M}}{2.0\text{ M}} = 1.147\text{ V}$$

c. Writing the equation for the net reaction and remembering that the anode is on the left side of the diagram:

$$Cu(s) + Cl_2(g) \leftrightarrow Cu^{2+}(aq) + 2\,Cl^-(aq)$$

$$E^o_{cell} = E^o_{oxid} + E^o_{red} = -0.337\text{ V} + 1.358\text{ V} = 1.021\text{ V}$$

$$E_{cell} = 1.021\text{ V} - (0.0592/2)\log Q$$

$$= 1.021\text{ V} - (0.0296)\log\frac{[Cu^{2+}][Cl^-]^2}{P_{Cl_2}}$$

$$= 1.021\text{ V} - (0.0296)\log\frac{(1.0\text{M})(0.25\text{M})^2}{0.50\text{atm}} = 1.048\text{ V}$$

9. Concentration cells

From the Nernst equation, we can see that a concentration difference between identical electrodes will give rise to a cell voltage. Such cells are called concentration cells. The value of E^o_{cell} will be zero for identical electrodes.

We can write the expression for E_{cell} for a concentration cell using a standard hydrogen electrode.

The expression is $E_{cell} = 0.0592\text{ pH}$

10. Batteries

A battery is an assembly of two or more voltaic cells connected together.

A dry cell is the type of battery used in most flashlights and electronic devices. Cell reactions are irreversible. The anode is a zinc container and the cathode is a carbon rod in contact with $MnO_2(s)$. The electrolyte is a moist paste of NH_4Cl and $ZnCl_2$. The cell diagram is:

$Zn(s)| ZnCl_2(aq), NH_4Cl(aq), || Mn_2O_3(s) | MnO_2(s), C(s)$

A second type of battery is a lead acid storage battery. In this type of battery, the cell reaction can be reversed. The anodes are lead alloy, the cathodes are lead alloy containing lead dioxide and the electrolyte is dilute sulfuric acid.

The third type of cells are fuel cells. The reaction is fuel + oxygen --> oxidation (combustion) products.

11. Corrosion: metal loss through voltaic cells

Corrosion of iron occurs when oxidation and reduction occur at separate points on the metal. The net reaction is:

$$4\,Fe(s) + 3O_2(g) + 6\,H_2O(l) \longrightarrow 4Fe(OH)_3(s)$$

Corrosion can be prevented by coating the iron surface or by using of a more active metal as a sacrificial anode. The more active metal acts as an anode and protects the iron because oxidation occurs at the active metal.

12. Electrolysis

Electrolysis is a process of using electricity to produce a nonspontaneous chemical change.

Example 18.8. Write plausible half-equations and a net equation for the electrolysis of KI(aq).

Solution: The principal species in solution are K^+ and I^- ions and H_2O molecules. From Example 18.8 we know that the reduction half-reaction will be:

$$2\,H_2O + 2\,e^- \longrightarrow H_2(g) + 2\,OH^-(aq) \quad E^o_{red} = -0.828\text{ V}$$

The possible oxidation half-reactions and their E^o_{oxid} values are:

$$2\,I^-(aq) \longrightarrow I_2 + 2\,e^- \qquad E^o_{oxid} = -0.535\text{ V}$$
$$2\,H_2O \longrightarrow O_2(g) + 4\,H^+(aq) + 4\,e^- \qquad E^o_{oxid} = -1.229\text{ V}$$

The oxidation of I^- will predominate and the net electrolysis reaction will be

$$2\,I^-(aq) + 2\,H_2O \longrightarrow I_2(l) + H_2(g) + 2\,OH^-(aq)$$

$$E^o_{cell} = -0.828\text{ V} + (-0.535\text{ V}) = -1.363\text{ V}$$

Example 18.9. Write plausible half-equations and a net equation for the electrolysis of $Na_2SO_4(aq)$.

Solution: The principal ions in solution are Na^+ and SO_4^{2-}. H_2O is also present and may be oxidized or reduced. Look first at possible reduction half-reactions:

$$2\,H_2O + 2\,e^- \longrightarrow H_2(g) + 2\,OH^-(aq) \qquad E^o_{red} = -0.828\text{ V}$$

$$Na^+ + e^- \longrightarrow Na(s) \qquad E^o_{red} = -2.713\text{ V}$$

The first equation is the most favorable reduction half-reaction.

The possible oxidation half-reactions are:

$2\ SO_4^{2-} \text{ --> } 2\ e^- + S_2O_8^{2-}$ $\qquad E^o_{oxid} = -2.01$ V

$2\ H_2O \text{ --> } O_2 + 4\ H^+ + 4\ e^-$ $\qquad E^o_{oxid} = -1.229$ V

The second equation is the most favorable oxidation half-reaction. The net equation is the sum of the oxidation and reduction half-reactions. To balance the electrons, the reduction half-reaction must be multiplied by two:

Reduction:

$2\ H_2O + 2\ e^- \text{ --> } H_2(g) + 2\ OH^-(aq)$

Oxidation:

$2\ H_2O \text{ --> } O_2 + 4\ H^+ + 4\ e^-$

Canceling species that appear on both sides of the equation leads to:

$2\ H_2O \text{ --> } 2\ H_2(g) + O_2(g)$

13. Quantitative electrolysis

The quantity of reactant consumed is related to the molar mass of the substance, quantity of electric charge used and the number of electrons transferred in the electrode reaction.

Example 18.10. For how many minutes must the electrolysis of a solution of $CuSO_4(aq)$ be carried out with a current of 2.25 A to deposit 1.00 g of Cu(s) at the cathode?

Solution: In this case, we know the amount of product and need to calculate the moles of electrons, convert the moles of electrons to charge and then determine the time. First, write the cathode reaction:

$Cu^{2+}(aq) + 2\ e^- \text{ --> } Cu(s)$

Calculate the amount in moles of copper needed to produce 1.00 g:

1.00 g x 1 mol/63.55 g = 0.0157 mol Cu(s)

We need 2 mol e^- for each 1 mol of Cu(s) so the amount of e^- = 2 x 0.0157 mol = 0.0314 mol.

Converting the moles of e^- to charge:

0.0314 mol e^- x 96,485 C/mol = 3.03×10^3 C

1 A = 1 C/s

The current is 2.25 A. The time will be

s = Coulombs/Amperes = 3.03×10^3 C/(2.25 Cs^{-1}) = 1.35×10^3 s.

Quiz A

1. For the reaction:

$H_2(g) + 2\ Fe^{3+}(aq) \text{ --> } 2\ H^+(aq) + 2\ Fe^{2+}(aq)$ $\qquad E^o_{cell} = +0.771$ V.

What is the value of E^{o}_{red} for the reaction:

$2\ Fe^{3+}(aq) + 2\ e^{-} \text{-->} 2\ Fe^{2+}(aq)$?

2. The anode in an electrochemical cell
   a. is always solid
   b. is where reduction occurs
   c. is where oxidation occurs
   d. is unchanged in the reaction
   e. none of these

3. If an electrochemical cell has $E^{o}_{cell} < 0$
   a. The cell reaction is spontaneous in the direction written
   b. the cell reaction cannot occur
   c. the cell reaction is spontaneous in the opposite direction
   d. none of the above

4. Write the equations for the half-reactions for the cell diagrammed below:

   $Zn(s)\ |\ Zn^{2+}(aq)\ ||\ Cu^{2+}(aq)\ |\ Cu(s)$

5. One ampere of current is defined as
   a. one mole of e^{-}
   b. 1 C/s
   c. E^{o}_{cell} for a standard hydrogen electrode
   d. 0.0592 C/s

6. Write a cell diagram for the voltaic cell that has the net reaction:

   $Cr_2O_7^{2-}(aq) + 6\ Fe^{2+}(aq) + 14\ H^{+}(aq) \text{-->} 2\ Cr^{3+}(aq) + 6\ Fe^{3+}(aq) + 7\ H_2O(l)$

7. What is the potential assigned to the standard hydrogen electrode?

8. Predict whether the oxidation of $Cl^{-}(aq)$ to $Cl_2(g)$ by I_2 will occur to any appreciable extent.

9. Write the cell reaction and E^{o}_{cell} of the spontaneous cell made from Ag/Ag^{+} (-0.80 V) and Cl^{-}/Cl_2(-1.36 V).

10. Describe, using equations as needed, the corrosion of metallic iron.

Quiz B

1. The cathode in an electrochemical cell is
   a. Pt
   b. where reduction occurs
   c. where oxidation occurs

d. unchanged in the reaction

e. none of these

2. For the cell diagrammed below which is the anode and which is the cathode?

   $Fe(s) | Fe^{2+}(aq) || Cu^{2+}(aq) | Cu(s)$

3. For the cell diagrammed in question 2, write the half-reactions.
4. Which of the following metals will displace sodium ions from solution?

   a. Ag^{+}

   b. Fe^{3+}

   c. Li^{+}

   d. none of these

   e. all of these

5. For the cell

   $Zn(s) | Zn^{2+}(aq)\ (2.0\ M) || Cu^{2+}(aq)\ (0.05\ M) | Cu(s)$

   a. $E_{cell} = E^{o}_{cell}$

   b. $E_{cell} > E^{o}_{cell}$

   c. $E_{cell} < E^{o}_{cell}$

   d. cannot predict

6. Describe a galvanic or voltaic cell.
7. The number of coulombs of charge per mole of electrons is one way to express:

   a. Avagadro's number

   b. Faraday's constant

   c. the Nernst equation

   d. the cell potential of the hydrogen electrode

   e. none of the above

8. Give the equation to determine the free energy from the cell potential.
9. What is a battery?
10. Describe the principle behind cathodic protection.

Quiz C

True or False

1. In an electrolytic cell, a spontaneous chemical reaction generates an electric current.
2. The anode is the electrode at which oxidation takes place.
3. Salt bridges maintain electric neutrality by a flow of ions.
4. The potential of a single electrode can be measured.
5. A positive value of E^{0} corresponds to $K_{eq} > 1$.

6. Corrosion occurs with reduction of metal.
7. The amount of substance produced at an electrode by electrolysis depends on the quantity of charge passed through the cell.
8. The cell potential of a galvanic cell is positive.
9. The oxidizing agent in a redox reaction is the species that causes oxidation to occur and is itself reduced.

Self Test

Fill in the blank

1. ________________ is the electrode at which reduction takes place.
2. ________________ is the oxidation of, and surface loss, of metal.
3. The standard cell potential is the cell potential when both reactants and products are ________________.
4. Whenever the direction of a half-reaction is reversed, the sign of E^0 is________________.
5. An oxidizing agent can oxidize any reducing agent that lies ________________ in the table of standard reduction potentials.
6. Cell potentials depend on ____________ and ______________ of the reaction mixture.
7. The ________________ allows us to calculate cell potentials under non-standard-state conditions.
8. In writing cell notation, the ______________ is always written on the right.

Problems/short answer

9. Write the half-reactions and determine whether the following reaction will occur spontaneously (use the table of standard reduction potentials):

   $SO_4^{2-}(aq) + 4H^+ (aq) + 2I^- (aq) \rightarrow H_2SO_3(aq) + I_2(s) + H_2O (l)$

10. Calculate ΔG^0 and K for the following reaction:

    $O_2(g) + 4H^+(aq) + 4Fe^{2+}(aq) \rightarrow 2H_2O(l) + 4Fe^{3+}$

11. What is the maximum mass, in grams, of Cl_2 produced in the electrolysis of molten NaCl by a current of 5.00 A at sufficient voltage for 25 min.?
12. Describe several types of batteries. Why can some batteries be recharged and others cannot?
13. Arrange the following species in order of increasing strength as reducing agents: H_2, Zn, Ni, F_2.
14. What change in emf of a hydrogen-copper cell occurs when NaOH is added to the solution in the hydrogen half-cell.?

CHAPTER 19

Nuclear Chemistry

Radioactivity is the spontaneous decay of the nuclei of certain atoms, accompanied by the emission of subatomic particles and/or electromagnetic radiation. This chapter will be concerned with the applications of radioactivity to chemistry, the life sciences and medicine as well as nuclear energy and ionizing radiation.

Chapter Objectives:

1. Describe the different types of radioactive decay and write nuclear equations to represent the decay.
2. Know that there are naturally occurring radioactive elements.
3. Determine the amount of radioactive material present from the radioactive decay law.
4. Describe how carbon-14 decay is used to determine the age of organic matter
5. Know how transuranium elements are formed
6. Describe the relationship between the number of neutrons and protons and nuclear stability.
7. Express the relationship between mass and energy and calculate the mass and energy change for nuclear reactions.
8. Define nuclear binding energy
9. Define nuclear fission and nuclear fusion.
10. Describe a nuclear reactor.
11. Describe ionizing radiation and its effects on matter.
12. List some of the used or radioactive nuclides.

Chapter Summary

1. Radioactive Decay

There are five types of radioactive decay and these can be represented by nuclear equations.

In a nuclear equation, particles are represented as ${}^{A}_{Z}X$ where Z is the atomic number, A is the mass number, and X is the element. The two sides of a nuclear equation must have the same totals of atomic numbers and mass numbers.

Alpha particles are extremely fast-moving, excited, doubly-ionized helium nucleus and contain two protons and two neutrons. They have a 2+ charge. Beta particles are even faster moving, highly excited electrons and have a charge of –1 and essentially zero mass. The mass of a nucleus when it emits a beta particle is unchanged. Gamma rays are an extremely energetic form of electromagnetic wave radiation, with wavelengths on the order of the size of nuclei, and are not considered as matter. They are emitted as a means of reaching a lower energy state. Positrons are positively charged particles having the same mass as an electron. Electron capture occurs as the nucleus absorbs an electron from the innermost (n = 1) energy level. It is accomplanied by release of x-rays and other electromagnetic radiation as electrons in higher n levels "cascade" down, refilling the unfilled n = 1 level.

Example 19.1. Write the reaction for ^{14}N reacting with a neutron to form ^{14}C and a product. Identify the product.

Solution: The equation will be

$$^{14}_{7}N + ^{1}_{0}n \rightarrow ^{14}_{6}C + ?$$

The sum of the mass numbers of reactants must equal the sum of the mass numbers of products. Therefore the unknown particle must have a mass of +1. The sum of the nuclear charges on both sides of the equation must also be the same. The particle must have a charge of 1, making it a proton. The equation is then

$$^{14}_{7}N + ^{1}_{0}n \rightarrow ^{14}_{6}C + ^{1}_{1}p$$

2. Naturally occurring radioactivity

There are a number of naturally occurring radioactive elements, such as carbon-14, U-238 and others.

3. Radioactive decay law

The radioactive decay law states that the rate of disintegration of a radioactive nuclide is directly proportional to the number of atoms present at that instant in time.

Example 19.2. The half-life of plutonium-239 is 2.411×10^4 y. How long would it take for a sample of plutonium-239 to decay to 1.00% of its present activity?

Solution: Solve for the rate constant of the reaction, λ, from the half-life and use the integrated rate law to determine the time. Solving for λ:

$$\lambda = \frac{0.693}{2.411 \times 10^4 \text{ y}} = 2.87 \times 10^{-5} \text{ y}^{-1}.$$

Using the integrated rate law: $\ln \frac{N_t}{N_0} = -\lambda t$ where $N_t = 0.0100\ N_0$. Substituting into the rate law:

$$\ln \frac{0.0100\ N_0}{N_0} = -2.87 \times 10^{-5} \text{ y}^{-1} \times t$$

Solving for t, the time required:

$$t = \frac{\ln 0.0100}{-2.87 \times 10^{-5} \text{ y}^{-1}} = 1.60 \times 10^5 \text{ y}$$

4. Carbon-14 decay and the age of organic matter.

In living organisms, the naturally occuring ^{14}C is both decaying and being replenished from the atmosphere. This results in a fairly constant level of ^{14}C so long as the organism is alive. When the organism dies, no replenishment occurs. By measuring the remaining ^{14}C activity and extrapolating the original mass of ^{14}C to be present, the time elapsed since death can be estimated. For the method to be useful, the time elapsed must be on the order of the ^{14}C half-life; a significant fraction of, or a small multiple of, 5715 years.

Exercise (text) 19.4B. Tritium (^3H), a beta-emitting hydrogen nuclide, can be used to determine the age of items up to about 100 years. A sample of brandy, stated to be 25 years old and offered for sale at a premium price, has tritium with half the activity of that found in new brandy. Is the claimed age of the beverage authentic? Use data from Table 19.2 and assume the natural abundance of tritium is a fixed quantity.

Solution: The half-life of tritium is 12.26 years. A common sense approach says that if the activity is 1/2 that of new brandy, one half-life has passed, that is, 12.26 years. The brandy is therefore about half as old as claimed. To work this out exactly, the activity is half of the original sample. Solve this by using the radioactive decay law, the integrated rate expression and determining the decay constant. Using the radioactive decay law:

$$N_o = A_o/\lambda \text{ and } N_t = A_t/\lambda$$

Using the integrated rate law:

$$\ln (N_t/N_o) = \ln (A_t/A_o) = \ln (0.5A_o/A_o) = \ln (0.5) = -\lambda t$$

and

$$\lambda = 0.693/t_{1/2} = 0.693/12.26 \text{ y} = 0.0565 \text{ y}^{-1}.$$

Substituting into the integrated rate equation:

$$\ln (0.5) = -0.0565 \text{ y}^{-1} \times t$$

$$t = 12.3 \text{ y}$$

The brandy is only about half as old as advertised.

5. Transuranium elements

1. Transuranium elements are those with $Z > 92$. The first of these was synthesized in 1940. This was accomplished by bombarding uranium-238 with neutrons.

6. Nuclear stability

1. Nuclei with "magic numbers" of protons or neutrons are more stable. The "magic numbers" are 2, 8, 20, 28, 50, 82, 126. A crucial factor in stability of the nucleus is the ratio of neutrons/protons. Figure 19.4 shows the belt of stability of neuclei. All nuclides falling outside this belt will decay by a mode that brings the nuclides formed into the belt.

2. In general, nuclei with even numbers of both protons and neutrons are more stable than those with odd numbers of these particles. Nuclei with even numbers of either neutrons or protons are more stable than those with odd numbers of both. All isotopes of elements after bismuth ($Z = 83$) are radioactive.

Example 19.3. Using the stability rules, which one of the following isotopes is most stable?

^{3}H ^{16}O $^{222}_{86}Rn$ $^{98}_{43}Tc$

Solution: $^{222}_{86}Rn$ will be least stable because all elements after bismuth are radioactive. $^{98}_{43}Tc$ will be unstable because it has an odd number of both protons and neutrons. ^{16}O will be most stable because it has an even number of neutrons and protons.

7. Energy of Nuclear Reactions

Einstein derived the relationship equating mass and energy

$$E = mc^2$$

In a typical nuclear reaction, a small quantity of matter is destroyed and replaced by a corresponding quantity of energy. Nuclear energies are typically in megavolts, MeV. An electronvolt is the energy that an electron acquires as it moves through a potential difference of one volt.

Example 19.4. Calculate the energy required for the process:

$^{3}_{2}He \rightarrow 2\ ^{1}_{1}p + ^{1}_{0}n$ given the following atomic masses: proton mass = 1.007825 u, neutron mass = 1.008665 u, and He atomic mass = 3.01603 u.

Solution: First calculate the difference in atomic mass and then use the relationship between MeV and u developed in the text (931.5 MeV/u). You can, of course also convert u to kg and use the Einstein equation to determine the energy. This is how the conversion factor above was arrived at.

The difference in atomic mass is

$$\Delta m = 2(\text{proton mass}) + (\text{neutron mass}) - {}^{3}_{2}He(\text{nuclear mass})$$

$$= (2)(1.007825\ u) + 1.008665\ u - 3.01603\ u = 8.29 \times 10^{-3}\ u$$

Energy is

$$E = 8.29 \times 10^{-3}\ u \times 931.5\ MeV/u = 7.72\ MeV$$

8. Nuclear Binding Energy

The energy released in forming a nucleus from protons and neutrons is the nuclear binding energy. The nuclear binding energy can be calculated from the mass differences, and in fact, Example 19.4 above calculates the nuclear binding energy.

9. Nuclear Fission and Nuclear Fusion

The process of combining light nuclei into heavier ones is nuclear fusion. Nuclear fission is the breakup on heavy nucleus into two different lighter nuclei and 2-4 fast neutrons. Nuclear reactors produce energy by controlled nuclear fission. A critical mass is necessary to sustain the chain reaction involved in a nuclear reactor. Water is used in a nuclear reactor as a moderator to slow neutrons down and as a heat-transfer medium.

10. Effects of Radiation on Matter

Ionizing radiation can cause tissue damage in living organisms. The potential harm from exposure to radioactive material depends on several factors, including the penetrating power of the radiation and whether the radiation is inside the body or the body is exposed from outside. The length of exposure can also have a significant impact.

11. Applications of Radioactive Nuclides

Some of the applications of radioactive nuclides include: cancer treatment, sterilization of medical equipment, sterilization of insects, radioactive tracers, and, in research, the determination of reaction mechanisms.

Quiz A

True or False

1. The sum of the mass numbers of the reactants in a nuclear reaction is not equal to the sum of the mass numbers of the products because of the nuclear binding energy.
2. An alpha particle is a ^{4}He nucleus.
3. A positron has a +1 charge and 1 u mass.
4. A beta particle has a –1 charge and the mass of an electron.
5. Nuclei with odd numbers of protons and neutrons are the most stable.
6. Magic numbers refer to the number of isotopes that can be formed.
7. There are no naturally occurring radioactive elements.
8. Radioactive decay rates obey first order kinetics.
9. The energy released by nuclear fission is large compared to that in ordinary chemical reactions.
10. When ^{235}U nuclei undergo neutron-induced fission, they all split into the same two nuclei plus three neutrons.

Quiz B

Multiple Choice

1. The following device is used to measure radiation:
   a. nuclear reactor
   b. x-ray spectrometer
   c. geiger counter
   d. mercury battery
2. Magic numbers of neutrons and protons:
   a. predict fusion products
   b. undergo rapid decay

c. occur in very large isotopes

d. give rise to particularly stable nuclei

3. A loss of mass occurs when protons and neutrons combine to form a nucleus. The lost mass:

   a. is never regained

   b. is converted into energy

   c. travels at the speed of light

   d. produces a chemical reaction

4. The relationship between the energy change of a nuclear process and the corresponding mass change:

   a. is inversely proportional

   b. was derived by Einstein

   c. cannot be determined

   d. is related to the change in outer shell electrons

5. Transuranium elements:

   a. have been produced by nuclear bombardment

   b. are decay products of uranium-235

   c. are more stable than lower atomic number elements

   d. contain an equal number of neutrons and protons

6. Nuclear fission:

   a. is the fragmentation of heavy nuclei

   b. doesn't occur in the same way every time

   c. releases energy

   d. all of the above

   e. none of the above

7. The radioactive half-life:

   a. is the time required for 1/2 the number of radioactive nuclei present in a sample to decay

   b. depends on the size of the sample

   c. depends on the temperature

   d. all of the above

   e. none of the above

Self Test

Matching

1. critical mass — a. consists of two protons and two neutrons. It is emitted by the nuclei of some radioactive elements

2. beta particle — b. the splitting of a large unstable nucleus into two lighter fragments and two or more neutrons.
3. half-life — c. the joining together or fusing of lighter nuclei into a heavier one
4. alpha particle — d. identical to an electron and is emitted by the nuclei of certain radioactive elements
5. gamma ray — e. the minimum mass of a fissionable element that must be present to sustain a chain reaction
6. nuclear fusion — f. a highly penetrating form of radiation emitted by the nuclei of some radioactive elements as they undergo decay.
7. nuclear fission — g. the time in which one-half of the atoms of a radioisotope disintegrate.

Short Answer/Problems

8. Balance the following nuclear reactions:

   a. ${}^{104}_{47}Ag \rightarrow {}^{0}_{1}e +$

   b. ${}^{104}_{48}Cd \rightarrow {}^{104}_{47}Ag +$

9. Write nuclear equations for:

   a. positron emission by ${}^{74}_{35}Br$

   a. beta emission by ${}^{73}_{31}Ga$

10. What is the age of a piece of wood that shows an average of 4.0 disintegrations per minute per gram of carbon? Carbon in living organisms undergoes an average of 15.3 dpm/g, and the half-life of ${}^{14}C$ is 5715 y.

CHAPTER 20

The s-Block Elements

The s-Block elements have valence electrons in the s subshell only and include hydrogen, the alkali metals (Group 1A), and the alkaline earth metals (Group 2A).

Chapter Objectives: You will be able to:

1. Describe some of the properties of hydrogen, including electron configuration and abundance.
2. Know the reactions that are the principal source of hydrogen gas commercially.
3. Describe the reactions of hydrogen.
4. Know the important commercial uses of hydrogen.
5. Explain the use of hydrogen as a reducing agent.
6. Describe the advantages and disadvantages of using hydrogen as a commercial fuel.
7. Know the electron configuration and properties of alkali metals.
8. Write the reactions of Group 1A metals.
9. Name some active metals and write the reactions of the active metals.
10. List several uses of alkali metal compounds and know how to write plausible chemical equations for conversion of compounds.
11. Know the characteristics of Group 2A metals.
12. Describe the Dow process.
13. Discuss the typical reactions of the alkaline earth metals.
14. Describe the important compounds of magnesium and calcium.
15. Define hydrate and give a general formula for alkaline earth metal hydrates.
16. Explain the process of ion exchange and its use in water softening.

Chapter Summary

1. Hydrogen

A hydrogen atom contains one proton and one electron in the $1s^1$ configuration. Hydrogen occurs as a diatomic molecule, H_2. In terms of numbers of atoms, H is the third most abundant element in the Earth's crust if we include the oceans and atmosphere.

Hydrogen is obtained from the reaction of its compounds.

The reactions are:

$$C(s) + H_2O(g) \xrightarrow{1000\ ^\circ C} CO(g) + H_2(g)$$

$$CO(g) + H_2O(g) \xrightarrow{1000\ ^\circ C,\ \text{catalyst}} CO_2(g) + H_2(g)$$

$$CH_4(g) + H_2O(g) \xrightarrow{1000\ ^\circ C,\ \text{catalyst}} CO(g) + 3\ H_2(g)$$

Example 20.1. What mass of $H_2(g)$ is obtainable from 1.00 kg of Zn(s)?

Solution: Use is the reaction of Zn with acid:

$$Zn(s) + 2\ H^+(aq) \text{ --> } Zn^{2+}(aq) + H_2(g)$$

One mole of H_2 is produced for each mole of Zn. Therefore:

$$1.00 \text{ kg} \times \frac{1.00 \times 10^3 \text{ g}}{1 \text{ kg}} \times \frac{\text{mol Zn}}{65.39 \text{ g}} \times \frac{1 \text{ mol } H_2}{1 \text{ mol Zn}} \times \frac{2.02 \text{ g}}{1 \text{ mol } H_2} = 30.9 \text{ g}$$

2. Reactions of Hydrogen

Hydrogen reacts with nonmetals to form binary compounds such as HCl.

$H_2(g) + Cl_2(g) \text{ --> } 2\ HCl(g)$

Hydrocarbons are compounds of carbon and hydrogen.

Hydrogen reacts with most active metals to form ionic hydrides.

Example 20.2. Write the equation for the reaction of hydrogen with potassium.

Solution: Potassium is a Group 1A element and reacts with hydrogen to form an ionic hydride.

$2\ K(s) + H_2(g) \text{ --> } 2\ KH(s)$

Example 20.3. Write the equation for the reaction of hydrogen with barium and subsequently the reaction with water.

Solution: Barium is an active metal and forms an ionic hydride. Ionic hydrides react vigorously with water to produce $H_2(g)$. The first reaction is:

$Ba(s) + H_2(g) \text{ --> } BaH_2(s)$

The subsequent reaction with water is:

$BaH_2(s) + 2\ H_2O \text{ --> } Ba^{2+}(aq) + 2\ OH^-(aq) + 2\ H_2(g)$

3. Commercial uses of hydrogen

The largest single use of hydrogen is in the production of ammonia.

$3\ H_2(g) + N_2(g) \text{ --> } 2\ NH_3(g)$

A second important use is in hydrogenation reactions, where hydrogen atoms are added to double or triple bonds to convert unsaturated hydrocarbons to saturated hydrocarbons.

Example 20.4. Write the equation for the hydrogenation reaction of propene (C_3H_6) to form propane (C_3H_8).

Propene contains a double bond to which hydrogen atoms can be added.

$$CH_3\text{-}CH{=}CH_2 + H_2 \xrightarrow{\text{catalyst}} CH_3\text{-}CH_2\text{-}CH_3$$

4. Hydrogen as a reducing agent

An example of hydrogen used as a reducing agent is the preparation of tungsten.

$$WO_3(s) + 3\ H_2(g) \xrightarrow{850\ ^{\circ}C} W(s) + 3\ H_2O\ (g)$$

Example 20.5. Write an equation to describe the preparation of solid nickel from nickel oxide.

Solution: Hydrogen can be used as a reducing agent. The reaction is:

$$NiO(s) + H_2(g) \text{ --> } Ni(s) + H_2O(g)$$

5. Hydrogen as a commercial fuel

Advantages of using hydrogen as a commercial fuel include high efficiency and low pollution. One major disadvantage is that hydrogen forms explosive mixtures on contact with oxygen.

6. Alkali metals

Alkali metals have the valence shell electron configuration of ns^1. They readily form +1 ions. Physical properties of group 1A metals are related to their large atomic radii. Alkali metals are very reactive, soft solids with low melting points and low densities.

Liquid sodium is used as a heat transfer medium in some types of nuclear reactors. Lithium is used in light weight batteries.

7. Properties and preparation of alkali metals

Many alkali metals are found as compounds such as NaCl. In order to produce an alkali metal from an alkali metal compound, the alkali metal ion must take on an electron. This is generally accomplished by electrolysis and is an important method of preparation of alkali metals.

$$2\ NaCl(l) \xrightarrow{\text{electrolysis}} 2\ Na(l) + Cl_2(g)$$

8. Reactions of Group 1A metals

Group 1A metals react directly with elements of Group 7A, the halogens to form ionic binary halides (MX).

$$2\ K(s) + Br_2(g) \text{ --> } 2\ KBr$$

Reaction with hydrogen:

$$2\ Na(s) + H_2(g) \text{ --> } 2\ NaH(s)$$

Reaction with excess oxygen:

$$4\ Li(s) + O_2(g) \text{ --> } 2\ Li_2O(s) \text{ (and some } Li_2O_2)$$

Reaction with water:

$$2\ Na(s) + 2\ H_2O(l) \text{ --> } 2\ NaOH(aq) + H_2(g)$$

9. Reactions of lithium, sodium and potassium

Lithium, sodium and potassium are active metals in that they will displace $H_2(g)$ from water.

$$2\ M(s) + 2\ H_2O\ (l) \text{ --> } 2\ MOH(aq) + H_2(g)$$

10. Uses of alkali metals

The uses of alkali metal compounds are outlined in Table 20.3 of the text. The conversions are outlined in Figure 20.5. Some of the reactions are given in the examples below.

Example 20.6. Write the plausible chemical equations for each of the following conversions outlined in Figure 20.5.

a. NaCl to NaH **b.** NaCl to NaOCl

Solution:

a. Electrolysis of liquid NaCl produces Na metal. In the presence of heat and $H_2(g)$, NaH is formed.

$$2\ NaCl\ (l) \xrightarrow{\text{electrolysis}} 2Na(l) + Cl_2(g)$$

$$2\ Na(l) + H_2(g) \text{ ---> } 2\ NaH$$

b. Electrolysis of aqueous NaCl produces NaOH. Reaction with Cl_2 then produces NaOCl.

$$2\ NaCl(aq) + 2H_2O \xrightarrow{\text{electrolysis}} 2\ NaOH(aq) + Cl_2 + H_2$$

$$2\ NaOH + Cl_2 \text{ ---> } NaOCl + NaCl + H_2O$$

Example 20.7. Write plausible chemical equations for the conversion of Na_2CO_3 to Na_2SO_3.

Solution: First convert Na_2CO_3 to NaOH, then convert NaOH to Na_2SO_3. There are other possible conversion routes that can be used. This is a relatively direct one.

$$Na_2CO_3 + Ca(OH)_2 \text{ --> } 2\ NaOH + CaCO_3$$

$$SO_2 + 2\ NaOH \text{ --> } Na_2SO_3 + H_2O$$

The first member of a group of the periodic table often differs from the other group members. In the case of the Group 1A elements, lithium differs from some of the other members of the group, probably because of the high charge density (ratio of ionic charge to ionic radius). In some ways, lithium resembles Mg and its compounds. This is an example of a diagonal relationship.

11. Group 2A: Alkaline earth metals

The following generalizations apply to Group 2A metals with the exception of beryllium. The metals are soft, reactive (they give up two electrons), and form almost exclusively ionic compounds.

Example 20.8. Barium can be prepared by the reaction of BaO with aluminum at about 1800° C. The other product is Al_2O_3. Write a balanced equation for the reaction, including the physical state of each substance.

Solution: First write the reactants and products:

$BaO + Al \rightarrow Ba + Al_2O_3$

Balancing the equation gives:

$3\ BaO(s) + 2\ Al(l) \rightarrow 3\ Ba(g) + Al_2O_3(s)$

Use a handbook or other reference to look up the melting point and boiling point of each substance. BaO melts at 1923 °C and will be a solid. Ba boils at 1140 °C so it will be a gas. Al melts at 660 °C and boils at 2467 °C, so it will be a liquid. Al_2O_3 melts at 2045 °C and is therefore solid.

12. The Dow process for preparation of Magnesium

The common source of magnesium is sea water. The Dow process (used for production of Mg) can be described as:

$CaCO_3(s) \xrightarrow{\Delta} CaO(s) + CO_2(g)$

$CaO(s) + H_2O \rightarrow Ca(OH)_2$

The hydroxide from $Ca(OH)_2$ reacts with Mg^{2+}.

$Mg^{2+}(aq) + Ca(OH)_2(aq) \rightarrow Mg(OH)_2(s) + Ca^{2+}(aq)$

$Mg(OH)_2(s) + 2\ HCl(aq) \rightarrow MgCl_2(aq) + 2\ H_2O(l)$

$MgCl_2 \xrightarrow{\text{electrolysis}} Mg(l) + Cl_2(g)$

13. Reactions of the alkaline earth metals

Reaction with halogens:

$M + X_2 \rightarrow MX_2$

Reaction with oxygen:

$2\ M + O_2 \rightarrow 2\ MO$

Reaction with nitrogen:

$3\ M + N_2 \rightarrow M_3N_2$

Example 20.9. Write chemical equations for the reaction:

a. Calcium with fluorine

b. Barium with oxygen

c. magnesium with nitrogen

The reaction of Ca with F_2 is:

$Ca + F_2$ --> CaF_2

The reaction of Ba with O_2 is

$2\ Ba + O_2$ --> $2\ BaO$

The reaction of Mg with N_2 is

$3\ Mg + N_2$ --> Mg_3N_2

14. Important compounds of magnesium and calcium

Limestone is calcium carbonate, $CaCO_3$, and it is an important building material and chemical raw material. The equations for the calcination of limestcne, hydration of quicklime and carbonation reaction are shown below.

Calcination is the decomposition of limestone by heating:

$CaCO_3(s)$ ---> $CaO(s) + CO_2(g)$

Quicklime is calcium oxide and the hydration reaction is

$CaO(s) + H_2O(s)$ --> $Ca(OH)_2(s)$

Calcium hydroxide is used in carbonation reactions:

$Ca(OH)_2(s) + CO_2(g)$ --> $CaCO_3(s) + H_2O(l)$

15. Hydrates

A hydrate is a compound that incorporates water molecules into its fundamental solid structure. Typical alkaline earth metal hydrates have the formula $MX_2 \cdot 6H_2O$ where M = Mg, Ca or Sr and X = Cl or Br.

Example 20.10. Which of the typical hydrates described by the formula $MX_2 \cdot 6H_2O$ has the greatest mass percent of water?

Solution: The typical hydrates are M = Mg, Ca or Sr and X = Cl or Br. We can, of course, add up the molecular weight of each of the possible compounds and calculate the mass percent water, but to estimate this we need to think about which compound will have the smallest molecular weight relative to $6H_2O$. This will give the highest mass percent of water. Therefore, M = Mg and X = Cl will have the highest mass percent water.

16. Hard water and water softening

Hard water is ground water that contains significant concentrations of ions from natural sources, mainly Ca^{2+} and Mg^{2+}. Ions, M^{2+}, are removed from hard water by formation of insoluble metal carbonates.

$$Na_2CO_3 + H_2O + Ca(HCO_3)_2 \text{ --> } 2\,Na^+ + 2\,HCO_3^- + CaCO_3(s)$$

Ion exchange is a means of softening hard water by using either a natural resin, such as zeolite, or a synthetic resin. When water containing divalent cations is passed through a zeolite resin, the dipositive cations are attracted and held to the resin. The Na^+ ions originally on the resin go into solution.

$$\text{-Z-}(Na)_2 + Ca^{2+}(aq) \text{ --> } \text{-Z-}Ca^{2+} + 2\,Na^+(aq)$$

The resin can be regenerated by using a concentrated NaCl solution to displace the Ca^{2+} ions.

A soap is the salt of a long-chain fatty acid. Compared to soaps,. detergents compared to soaps, do not leave precipitates in hard water.

Quiz A

1. Name the following compounds:

   a. Na_2CO_3

   b. CaO

   c. LiH

2. Complete the following equation:

   $CaO(s) + HCl(aq)$ -->

3. Complete the following equation:

   $2\,Mg + O_2$ -->

4. List three important Mg compounds and give one use of each.
5. What are the main sources of free hydrogen?
6. Give at least two ions typically involved in hard water.
7. How do the ionization energies and atomic radii of the Group 2A elements compare to those of the Group 1A elements?
8. Name the following compounds:

   a. $CaCO_3$

   b. $Ba(OH)_2$

Quiz B

1. Give the equation for the reaction of BaO with water.
2. Write the formula for calcium bromide hexahydrate.
3. Give the reaction for the calcination of limestone. Why is this reaction important?

4. Which have larger atomic radii, Group 1A metals or Group 2A metals?
5. What is meant by a diagonal relationship in the periodic table? Give an example.
6. Give the equation for the reaction of solid Zn with acid in aqueous solution.
7. Write the equation for the binary reaction of hydrogen and nitrogen.
8. What is a hydrogenation reaction?
9. Give the reaction of beryllium with water.
10. Write the chemical formula and chemical name for quick lime.

Quiz C

True or False

1. As a group, the alkali metals never occur naturally in the elemental form.
2. The alkali metals are highly electronegative.
3. Magnesium reacts rapidly with cold water.
4. Limestone is mostly calcium chloride.
5. The process for producing most free metals requires reduction of the metal ion.
6. Metallic sodium is obtained by electrolysis of molten sodium chloride.
7. The oxides of magnesium and calcium react with water to give basic solutions.
8. Water softening occurs by electrolysis of hard water.

Self Test

1. Write the reaction for the alkaline metals with halogens.
2. Describe the main process used for the production of magnesium.
3. Write the chemical equation for the reaction of HCl(aq) with Al(s).
4. Write the chemical equation for the reaction of Li(s) with water.
5. Describe the commercial uses of calcium.
6. What are the main ions present in hard water?
7. Write an equation for the calcination of limestone.
8. Write an equation for the production of hydrogen from methane.
9. Describe several commercial uses of magnesium.
10. Write an equation for a method of producing hydrogen other than from methane.

CHAPTER 21

The p-Block Elements

The p-block elements include the noble gases, all the nonmetals except hydrogen, all the metalloids, and a few metals, such as aluminum, tin and lead. This chapter describes the properties and reactions of these elements.

Chapter Objectives: You will be able to:

1. Know the properties and trends of the Group 3A elements.
2. Describe the characteristics of boron.
3. Know the formulas for borax and boric acid and describe the reaction of boric acid with water.
4. Describe the important properties of aluminum and the reaction for purification of aluminum from bauxite ore.
5. Describe the amphoteric nature of Al_2O_3.
6. Know the electron configurations and ions of gallium, indium and thallium and the importance of an inert pair.
7. Know some of the reactions of Groups 3A elements.
8. Describe the bonding of Group 4A elements.
9. Write reactions for the formation of a carbide.
10. Describe the differences and similarities of C and Si bonding in compounds.
11. List some of the similarities between the properties of tin and lead.
12. Describe the properties and electron configuration of the Group 5A elements.
13. Know the principal commercial source of nitrogen.
14. Describe the two allotropes of phosphorus.
15. Describe the reaction of phosphorus oxides with water.
16. Describe a few of the properties and uses of arsenic, antimony, and bismuth.
17. Know the characteristics of the Group 6A elements.
18. Know the characteristics of the Group 7A elements.
19. Describe the reactivity and commercial importance of the halogens.
20. Discuss why oxoanions are stronger oxidizing agents in acid than in base.
21. Describe the Group 8A elements.

Chapter Summary:

1. Properties and trends of Group 3A

The p-block elements are the elements of the periodic table Groups 3A through 8A. Group 3A elements have the valence shell electron configuration ns^2np^1. The group number indicates the maximum oxidation state of 3+. The elements at the top of the group, B and Al, form 3+ ions (except for a few complex solids). The oxidation number of +1 becomes more common moving down the group.

2. Boron

Boron is a nonmetal, which has a high first ionization energy and a high melting point. Many of the characteristics of boron compounds can be explained by the lack of an octet of electrons around the central boron atom. Boron forms an interesting and wide range of binary molecular compounds with hydrogen. Borane, BH_3, does not exist as a stable molecule; however, diborane, B_2H_6, does exist. Attempts to draw Lewis structures of diborane are complex because seven single bonds are needed (14 electrons) and only twelve valence electrons are available. Since B brings so few valence electrons, there are never any multiple bonds, nor any non-bonded electron pairs in these structures. Instead, the small total electron count requires that one two-electron, 3-center bond must be present for each neutral boron atom present. These may be of two types, B-H-B "bridge" interactions, and B-B-B "triangular" three-center interactions. Molecular orbital theory gives a clearer explanation of the bonding. Where three atoms combine, three molecular orbitals are formed. These three orbitals are a bonding (lowest energy), nonbonding (higher energy) and antibonding (highest energy). The two electrons of the three atoms will fill the bonding orbital and create a bond. Therefore, three atoms and two electrons form a bond, a "three center bond."

3. Borax and boric acid

The formula for borax is $Na_2B_4O_7 \cdot 10H_2O$. Borax is converted to boric acid by the reaction with sulfuric acid:

$$Na_2B_4O_7 \cdot 10H_2O + H_2SO_4 \text{ --> } 4\ B(OH)_3 + Na_2SO_4 + 5\ H_2O$$

Boric acid is $B(OH)_3$. We can understand why this is an acid by the reaction with water:

$$B(OH)_3(aq) + 2\ H_2O \text{ --> } H_3O^+(aq) + [B(OH)_4]^-(aq)$$

Exercise (text) 21.1A. Write an equation to represent the preparation of pure boron by the reduction of $BCl_3(g)$ with hydrogen gas.

Solution: The reduction half reaction must be:

$$B^{+3} + 3\ e^- \text{ --> } B.$$

The oxidation reaction is:

$$H_2(g) \text{ --> } 2\ H^+ + 2\ e^-$$

The combined reaction is:

$$2\ BCl_3 + 3\ H_2(g) \text{ --> } 2\ B + 6\ HCl(g)$$

4. Aluminum

Aluminum is the most important metal of Group 3A. Important properties of aluminum include: (1) it is a good reducing agent; (2) good electrical conductivity; (3) low density.

Aluminum sulfate is the most important industrial aluminum compound. It is used in water treatment.

The principal ore of aluminum is bauxite, which contains Al_2O_3 and SiO_2 and Fe_2O_3 are present as impurities. Treatment with concentrated NaOH dissolves Al_2O_3, leaves the impurities:

$$Al_2O_3(s) + 2\ OH^-(aq) + 3\ H_2O \text{ --> } 2\ [Al(OH)_4^-](aq)$$

Treatment with acid precipitates the aluminum as aluminum hydroxide:

$$[Al(OH_4)]^-(aq) + H_3O^+(aq) \text{ --> } Al(OH)_3(s) + 2\ H_2O$$

Heating results in the formation of the oxide

$$2\ Al(OH)_3(s) \text{ --> } Al_2O_3(s) + 3\ H_2O$$

and the final purification involves electrolysis between C electrodes:

$$2\ Al_2O_3(l) + 3\ C(s) \text{ --> } 4\ Al(l) + 3\ CO_2(g)$$

5. The amphoteric nature of Al_2O_3

Al_2O_3 dissolves in both acids and bases. Aluminum bonds to nonmetals have significant covalent character.

6. The electron configurations and ions of gallium, indium and thallium

Gallium, indium and thallium form 3+ and 1+ ions. An inert pair occurs when the valence shell pair of electrons (ns^2) is retained. This retention is common with post-transition elements. Since the inert pair is relatively stable, gallium will form a +1 ion.

7. Reactions of Groups 3A elements

Selected reactions are given below:

$2\ M + 3\ X_2 \text{ --> } 2\ MX_3$	X_2 = halogen, Tl gives TlX
$4\ M + 3\ O_2 \text{ --> } 2\ M_2O_3$	high temperature, Tl gives Tl_2O
$2\ M + 3\ S \text{ --> } M_2S_3$	Tl gives Tl_2S
$2\ M + 6\ H^+ \text{ --> } 2\ M^{3+} + 3\ H_2$	M = Al, Ga or In
$2M + 6\ H_2O \text{ --> } 2\ M(OH)_3^- + 3\ H_2$	M = Al or Ga

8. Group 4A elements

C, Si, Ge, Sn, and Pb have valence electron configurations of ns^2np^2. C shares the four electrons for covalent bonding in almost all of its compounds. Si and Ge also form covalent bonds for the most part. They are semiconductors. Sn and Pb form 2+ and 4+ ions.

9. Carbides

Reactions for the formation of a carbide, carbon disulfide from methane and sulfur, and carbon tetrachloride from methane and chlorine are given below.

Carbides are a combination of carbon and a metal. For active metals, the carbides are ionic. An example of carbide formation is the reaction of calcium oxide with carbon:

$CaO(s) + 3\ C(s) \dashrightarrow CaC_2(s) + CO(g)$

the reaction of methane to form carbon disulfide:

$CH_4(g) + 4\ S(g) \dashrightarrow CS_2(l) + 2\ H_2S(g)$

and the reaction of methane with chlorine:

$CH_4(g) + 4\ Cl_2(g) \dashrightarrow CCl_4(l) + 4\ HCl(g)$

10. C and Si bonding in compounds.

Silicates are the predominant naturally occurring compounds of silicon. Si-O-Si bonding in silicates is among the strongest known, and silicates make up 70% of the rocks and minerals in the earth's crust. Direct Si-Si bonding in organo-silicon compounds is generally weaker than the corresponding C-C bonding in organic C compounds. Because of the low activation energy of silicon chain and ring compounds they are less stable than carbon compounds. Si compounds, like C compounds, generally form four bonds. Si can form organosilicon compounds similar to, although less stable than, those of organic C compounds.

11. Properties of tin and lead

Tin and lead are soft, malleable and melt at relatively low temperatures. Their ionization energies and standard electrode potentials for reduction are similar.

12. Properties and electron configuration of the Group 5A elements

Group 5A: N, P, As, Sb, and Bi display properties that range from nonmetallic to metallic. All have valence shell electron configuration ns^2np^3. In a few cases they can gain three electrons to form a 3- ion. More often though, they share electrons to acquire an ns^2np^6 configuration. As we move down the group the following occur: increasing atomic size, decreasing ionization energy, and decreasing electronegativity.

13. Nitrogen compounds

Ammonia, a weak base, is the principal commercial source of nitrogen compounds. Hydrazine, NH_2NH_2, behaves much like two linked ammonia molecules. It reacts with water as a weak base that can ionize in two steps:

$NH_2NH_2(aq) + H_2O \dashrightarrow NH_2NH_3^+(aq) + OH^-(aq)$

$NH_2NH_3^+(aq) + H_2O \dashrightarrow NH_3NH_3^{2+}(aq) + OH^-(aq)$

The ionized form of hydrazine reacts with oxygen.

$NH_2NH_3^+(aq) + O_2(aq) \text{-->} 2\ H_2O + H^+(aq) + N_2(g)$

The direct reaction of liquid hydrazine with oxygen gas can be used for rocket propulsion.

$N_2H_4(l) + O_2(g) \text{-->} N_2(g) + 2\ H_2O(g)$

14. Allotropes of phosphorus

The two best-known allotropes of phosphorus are white phosphorus and red phosphorus. White phosphorus exists as discrete P_4 molecules and has a low melting point. It is soluble in nonpolar solvents. It is highly reactive due to the strained bond angles of 60°. Red phosphorus has a polymeric structure.

15. Reaction of phosphorus oxides with water

Phosphorus oxides react with water to form H_3PO_3 or H_3PO_4.

$P_4O_6(s) + 6\ H_2O(l) \text{-->} 4\ H_3PO_3$

$P_4O_{10}(s) + 6\ H_2O(l) \text{-->} 4\ H_3PO_4(aq)$

Phosphorus compounds have many uses, including fertilizer, laboratory drying agents, cleaning agents, commercial manufacture of cement, and soft drinks.

16. Some uses of arsenic, antimony, and bismuth

These three members of Group 5A are more metallic than nonmetallic in character. They are used in alloys. Arsenic compounds are used in insecticides.

17. Know the characteristics of the Group 6A elements

The elements of Group 6A have the valence shell electron configuration ns^2np^4. O and S are nonmetallic; Se, Te, and Po exhibit some metallic properties. Selenium and tellurium are semiconductors, polonium is an electrical conductor as well as being radioactive. Se is a photoconductor, which means its electrical conductivity increases in the presence of light.

O and S are similar because they both form ionic compounds with active metals. They form some similar covalent compounds such as H_2S and H_2O.

One difference between H_2O and H_2S is that H_2O has a much higher boiling point than H_2S. This is due to the hydrogen bonding in water. Sulfur can also form expanded octets since it has 3d orbitals available. Some of these differences can be attributed to the importance of hydrogen bonding in O compounds

18. Group 7A elements

Group 7A elements have the valence shell electron configuration ns^2np^5. They gain electrons easily and lose electrons only with difficulty. The electronegativity decreases and atomic radius increases going down the group.

19. Reactivity and commercial importance of the halogens

F is the most reactive of the halogens. The hydrogen halides form acids in aqueous solution. Except for HF, these are strong acids.

Element	Preparation	Commercial uses
F	electrolysis of HF	Production of UF_6 for nuclear fuels SF_6 insulating gas, HF used to etch glass
Cl	electrolysis of NaCl	chlorinated organic compounds, bleach, inorganic compounds, water purification
Br and I	treatment of brine	organic Br used as pharmaceuticals, dyes, fumigants, fire extinguishers
I		used in photographic emulsions, catalysis and in medicine
HF		behaves anomolously (higher boiling point, weak acid).

That HF is a weaker acid than other halogen halides is expected based on bond length and bond dissociation energies. We would expect that the stronger bond, the weaker the acid. This is because a strong bond would be less likely to dissociate. HF, however, is a weaker acid than expected. Explanations of this behavior have to do with the tendency for hydrogen bonding in HF. HF ion pairs are held together by hydrogen bonds which keeps the concentration of free H_3O^+ lower than otherwise expected.

20. Oxoanions—stronger oxidizing agents in acid than in base

If we examine the half reaction for reduction of ClO_4^- to ClO_3^- in acid and in base:

$ClO_4^- + 2\,H^+ + 2\,e^- \rightarrow ClO_3^- + H_2O$ (acid) $E^o_{red} = +1.19$ V

$ClO_4^- + H_2O + 2\,e^- \rightarrow ClO_3^- + 2\,OH^-$ (base) $E^o_{red} = +0.36$ V

We see that in acid solution, where $[H^+]$ (the reactant) is increased, by Le Chatelier's principle, the forward reaction is favored. In basic solution, where OH^- is a product, the reverse reaction is favored.

21. The group 8A elements—the noble gases.

These elements are unreactive although a few unstable molecules containing Xe, Kr, and Rn, almost always bonded to F or O, have been prepared.

Quiz A

1. In aqueous solution, HF is a weak acid while HCl is a strong acid. This difference can be attributed to
   a. smaller mass of HF
   b. electron configuration of HF
   c. larger mass of HF

d. hydrogen bonding in HF

e. none of these

2. H_2O has a much higher boiling point than H_2S. Explain this difference.
3. Write the formula for borax.
4. Write the formula for hydrazine.
5. List two commercial uses of chlorine.
6. Write the equation for the reaction of Al with a strong acid.
7. Write the electron configurations for the two ions of gallium.
8. What important compound is obtained from calcium carbide?
9. What is the main commercial use of lead?
10. Define and give an example of an allotrope.

Quiz B

1. Bauxite is a mineral source of which important metal?
2. Which element is a photoconductor?
3. What is the main Group 4A element found in quartz?
4. Write the reaction for ionization of hydrazine in aqueous solution.
5. Which of the noble gases will react with fluorine?
   a. none will react
   b. xenon
   c. all will react
   d. sodium
   e. argon
6. Would you expect F_2 or Br_2 to be a better oxidizing agent?
7. Write the equation for the reaction of $MnO_2(s)$ with $HCl(aq)$.
8. Which element tends to form electron-deficient compounds?
9. Which of the following elements, O, Se or Po, is the most nonmetallic?

Quiz C

Fill in the blank

1. Two minerals which are sources of boron are ___________ and ______________.
2. Boric acid can be prepared by treating borax with ___________________________.
3. Boron chloride reacts with water to form ___________________________.
4. Covalent hydrides are usually _____________ in water.
5. A ________________ is an ionic form of carbon.
6. __________ is used in the Ostwald process to make nitric acid.

Short Answer

7. List some of the commercial uses of the halogens.
8. Contrast the covalent hydrides of carbon with those of silicon.
9. Define an interhalogen
10. What is an inert pair and which elements are likely to contain an inert pair?

Self Test

1. What is the primary characteristic of a p-block element?
2. Which of the noble gases form compounds?
3. Write an equation to represent the preparation of pure boron by the reduction of $BCl_3(g)$ with hydrogen gas.
4. Write an equation for the reaction of bauxite with NaOH.
5. Write the outer shell electron configuration of the Group 5A elements and describe the properties moving down the group.
6. Write chemical equations to show the preparation of the elemental forms of the halogens.
7. Describe the Ostwald process for the preparation of nitric acid.
8. Write an equation for the Haber process for the commercial production of ammonia. What is the major commercial use of ammonia?
9. Explain why HF has a normal boiling point intermediate between those of NH_3 and H_2O and why HF is a significantly weaker acid than HCl.
10. List the known oxidation numbers of nitrogen and give a compound or ion that exemplifies each.

CHAPTER 22

The d-block Elements and Coordination Chemistry

This chapter deals primarily with properties of the d-block elements, including their reactions and the formation of complex ions and coordination compounds that contain complex ions.

Chapter Objectives: You will be able to:

1. Characterize d-block elements of the fourth period.
2. Know some of the other general properties of transition metals.
3. Describe the characteristics and properties of the elements scandium through manganese.
4. Describe the properties of Fe, Co, and Ni.
5. Characterize the properties of Group 1B elements: Cu, Ag, and Au.
6. Describe superconductors.
7. Characterize the Group 2B elements: Zn, Cd and Hg.
8. Define coordination compounds and describe Werner's coordination theory.
9. Define coordination complex, ligands, coordination number and complex ion.
10. Know the rules for naming common complex ions and be able to determine the formula of a complex ion from the name.
11. Describe structural isomers.
12. Describe optical isomers.
13. Give a brief description of crystal field theory.
14. Explain why many complex ions are colored.
15. Describe applications of coordination chemistry.
16. Define chelation and describe some uses of chelates.

Chapter Summary

1. d-block elements of the fourth period

1. Electronegativity values indicate metal characteristics.
2. Ions are produced by the loss of valence shell electrons and one or more 3d electrons.
3. They are able to conduct electricity, have relatively high melting points and high densities.

2. General properties of transition metals

1. Between the fourth and fifth periods atomic radii increase as expected, but stay about the same or decrease between fifth and sixth periods. In the sixth period, the 4f subshell must fill before the 5d. The shapes of the orbitals are such that the screening of valence electrons is reduced and as a result the effective nuclear charge is greater than expected. This stronger attraction of valence electrons accounts for the fact that atomic size does not increase between the fifth and sixth period members of a group of transition elements.

2. The lanthanide series in the sixth period has an electron configuration with filling of the 4f subshell.
3. Heavier group members have a greater range of oxidation states and more covalent character to their bonding.

3. Elements scandium through manganese

Scandium forms a 3+ ion and reacts with either acidic or basic solutions. It reacts similarly to aluminum.

Titanium is the ninth most abundant element and the chief mineral sources are rutile and ilmenite. Low density, high structural strength, and corrosion resistance make it a very useful metal. The most important compound is TiO_2, used as paint pigment.

Example 22.1. Write plausible equations for the two-step "chloride" process for purifying rutile ore to obtain pure $TiO_2(s)$. In the first step, impure $TiO_2(s)$ is converted to $TiCl_4(g)$ at about 800 °C. In the second step, $TiCl_4(g)$ reacts at about 1200 °C with oxygen to produce $TiO_2(s)$ and chlorine.

Solution: For the first reaction:

$$TiO_2(s) + 2\ C(s) + 2\ Cl_2(g) \text{ ---> } TiCl_4(g) + 2\ CO(g)$$

For the second reaction:

$$TiCl_4(g) + O_2(g) \text{ --> } TiO_2(s) + 2\ Cl_2(g)$$

Vanadium is obtained mostly as a byproduct from production of other metals, such as in the extraction of uranium from carnotite. It is used chiefly as an alloy in steel and some of its compounds have catalytic activity. Vanadium has oxidation states of 2+, 3+, 4+, and 5+. These species have distinctive colors.

The only important source of chromium is chromite, $FeCr_2O_4$. Chromium is used as an alloy to produce stainless steel and is also used to plate onto other metals.

Chromate and dichromate ions participate in a reversible reaction:

$$2\ CrO_4^{2-}(aq) + 2\ H^+(aq) \text{ <--> } Cr_2O_7^{2-}(aq) + H_2O$$

Dichromate ion is an important oxidizing agent with a large positive electrode potential. The main oxides are CrO_3, which is an acidic anhydride, and Cr_2O_3, which is amphoteric, as is its corresponding hydroxide, $Cr(OH)_3$.

The principal mineral containing manganese is pyrolusite, MnO_2. Manganese is used as an alloy for wear-resistant and shock-resistant steel. MnO_2-Fe_2O_3 is commonly reduced to obtain ferromanganese. The electron configuration of manganese is $[Ar]3d^5 4s^2$. Manganese displays a wide range of oxidation states; +2, +3, +4 and +7 are important.

Exercise (text) 22.2A. Show that $MnO_4^-(aq)$ is a good oxidizing agent in basic solution.

Solution: One reaction is the oxidation of $Cu(s) \text{ --> } Cu^{2+}(aq) + 2\ e^-$.
The half-equation for reduction is:

$MnO_4^- + 2\,H_2O + 3\,e^- \rightarrow MnO_2 + 4\,OH^-$ $\quad E^\circ_{red} = +0.60\ V$

The half-equation for oxidation is:

$Cu(s) \rightarrow Cu^{2+} + 2\,e^-$ $\quad E^\circ_{oxid} = -\ 0.337\ V$

The net reaction is:

$$2\,MnO_4^- + 3\,Cu(s) + 4\,H_2O \rightarrow 2\,MnO_2 + 2\,OH^- + 3\,Cu(OH)_2(s)$$

$$E^\circ_{cell} = E^\circ_{red} + E^\circ_{oxid} = 0.60\ V + (-0.337\ V) = 0.26\ V$$

4. Fe, Co, and Ni

Iron is the most important d-block metal and is widely distributed in the earth's crust. Co is found in the ore carrolite. Fe, Co and Ni are not only paramagnetic, but have a property called ferromagnetism where the magnetic moments of the individual atoms line up and produce domains. These domains line up when placed in a magnetic field and produce a strong magnetic effect.

These metals form metal carbonyls, neutral molecules with CO. These elements all form 2+ ions, but Fe forms a 3+ ion as well. The 3+ ion is particularly stable since it results in a half-filled d subshell. Complex ions can be formed by these metals.

5. Group 1B elements: Cu, Ag, and Au

These metal ions are easily reduced to the free metals. The metals are resistant to oxidation. The physical properties are characteristic of metals. Cu is used for electrical wiring; all of these metals are used for jewelry and decorative arts. These metals do not displace H^+ from solution.

6. Superconductors

Metals lose their resistance to the flow of electricity at low temperatures. Theoretically, this occurs at 0 °K, but for some metals this happens at liquid helium temperatures. These are superconductors.

7. Group 2B elements: Zn, Cd and Hg

This group of elements does not fit the usual definition of transition elements in that the d-subshells of their next-to-outermost electronic shells are filled.

Mercury differs from other members of this group in a number of ways: It is a liquid at room temperature; it will not displace $H_2(g)$ from acidic solutions; it can exist as +1 and forms a diatomic ion with a Hg-Hg covalent bond; it forms covalent compounds; most compounds are insoluble in water; and it has little tendency to oxidize.

Zinc and cadmium are used for coatings and alloys. When iron is coated with zinc it is said to be galvanized.

8. Coordination compounds and Werner's coordination theory

Coordination compounds consist of a "complex", with a central metal ion tightly surrounded by a number of "ligating" atoms to which the metal is covalently bonded. At greater distance, there may be additional, "non-ligating" ions to balance the overall net charge.
Werner's coordination theory stated that certain metal atoms have two types of valence or combining capacity: (1) a primary valence resulting from the loss of electrons; and (2) a secondary valence that is an attraction of a metal atom or ion for other groups that are attracted at specific positions around the metal.

9. Coordination complexes, ligands, coordination number and complex ions

Example 22.2. What are the coordination number and the oxidation state of the central atom in the complexes

a. $[Al(OH)_2(H_2O)_4]^+$

b. $[Fe(CN)_6]^{4-}$

Solution:

a. Six ligand groups -- 2 OH^- and 4 H_2O -- are attached to the Al central ion. The coordination number is six. This complex has a charge of +1. The H_2O ligands have no charge and the OH^- each have a -1 charge. The charge on the central ion must be x - 2 = +1 or x = +3. The central ion is Al^{3+} with an oxidation state of 3+.

b. There are six ligand groups-- CN^- --so the coordination number is six. Each CN group has a charge of 1- and the overall charge is 4-. The central atom has x - 6 = -4 or x = +2 charge. The oxidation state of Fe is 2+.

10. Naming common complex ions

The rules for naming complex ions are given in the text.

Example 22.3. Name the following:

a. $[Ag(NH_3)_2]^+$;

b. $[AuCl_4]^-$;

c. $[CoBr(NH_3)_5]Br_2$

Solution:

a. There are two ammonia molecules (ammine) as ligands. This gives a prefix of di. The ammonia molecules are not charged and the complex ion has a charge of 1+ so the Ag must have an oxidation state of +1. The complex is diamminesilver(I) ion. (Note the names for metals in Table 23.2.)

b. There are four (tetra) chloride(chloro) ions, each with a charge of 1-. The charge on the complex is 1-; therefore, gold has an oxidation state of 3+. The name is then tetrachloroaurate (III) ion.

c. There is one bromide ion (bromo) and five (penta) ammonia molecules (ammine) coordinated to the central cobalt ion. The bromide ion has a 1- charge and the ammonia molecules are neutral. The ion is associated with two bromide ions so the ion must have a 2+ charge. Because the bromide is 1- in the complex, the cobalt ion has a 3+ oxidation state. Name the compound by writing the ligands in alphabetical order: pentaamminebromocobalt(III) bromide.

Example 22.4. Write formulas for the following:

a. (tris)ethylenediaminecobalt(III) ion

b. diamminetetrachlorochromate (III) ion

c. dichlorobis(ethylenediamine)platinum(IV) sulfate

Solution.

a. The prefix tris is used for three since the ligand, ethylenediamine, already has a prefix di in the name. The abbreviation for ethylenediamine is en. This bidentate ligand has no charge so the charge on the ion will be 3+ from the cobalt. The formula is therefore $[Co(en)_3]^{3+}$.

b. The prefix di indicates two and ammine is NH_3 with no charge. The chloride ion has a -1 charge and there are four (tetra) chloride ions. Chromium has an oxidation state of +3, so the charge on the ion must be -4 +3 = -1. The formula is $[Cr(NH_3)_2Cl_4]^{1-}$.

c. There are two (di) chloride (chloro) ions each with a -1 oxidation state. The two (bis) ethylenediamine (en) ligands are uncharged. The oxidation state of the platinum is +4. The ion must be +2 and sulfate is –2, so the formula will be $[PtCl_2(en)_2]SO_4$.

11. Structural isomers

Structural isomers differ in the ligands that are attached to the central atom and/or the donor atoms through which they are bonded.

Exercise (text) 22.6B. Indicate whether different structural isomers are possible of the following:

a. $[Zn(H_2O)(NH_3)_3]^{2+}$

b. $[Cu(NH_3)_4][PtCl_4]$

Solution:

a. In this case, four ligands are attached tetrahedrally to the central Zn atom. The arrangement of these will not lead to different structural isomers.

b. In this case, we can write other possibilities for the distribution of ligands around the anion and cation, and this will lead to other isomers, for example, $[Pt(NH_3)_4][CuCl_4]$.

12. Optical isomers

Optical isomers are different geometric arrangements of atoms around the central atom such that the structure is not superimposable on its mirror image.

13. Crystal field theory

Crystal field theory views the attraction between the central atom or ion and its ligands as electrostatic. However, there are also repulsions between the ligand electrons and d electrons of the central atom. Common ligands have different abilities to split the d-orbital energy level.

Exercise (text) 22.8A and 22.8B. How many unpaired electrons would you expect to find in each of the following complex ions:

a. the octahedral complex ion, $[Co(CN)_6]^{3-}$

b. the tetrahedral complex ion $[NiCl_4]^{2-}$

Solution: We must (1) determine the number of 3d electrons in the central atom, (2) determine the energy difference between the d orbitals, and (3) distribute the electrons according to the aufbau principle.

a. $[Co(CN)_6]^{3-}$ The CN^- ions have a total charge of -6 for the six ligands. The complex has a charge of 3- so the cobalt ion has a 3+ charge. The atomic number of cobalt is 27, and the number of electrons in Co^{3+} is 24. The electron configuration is $[Ar]3d^6$ there are six d electrons to distribute. CN^- is a strong field ligand and, therefore, the energy splitting will be large. In Figure 23.11, the d orbital splitting has three lower energy orbitals and two higher energy orbitals. Because the splitting is large, we fill the lower energy orbitals first before the higher orbitals. The three lower energy orbitals will accommodate all six electrons as three pairs. There are no unpaired electrons.

b. $[NiCl_4]^{2-}$ The nickel ion has a 2+ charge since the four Cl^- ions each have a 1- charge and the complex has a 2- charge. Ni has 28 electrons and Ni^{2+} has 26 in an $[Ar]3d^8$ configuration. There are eight d electrons to distribute. Cl^- is a weak field ligand and so the splitting of the energy levels will be small. We will fill all five orbitals separately and then pair the last three electrons. This will leave two unpaired electrons.

14. Color and complex ions

Crystalline samples and solutions of metal complex ions often display colors. These colors arise when bands of visible light (where the corresponding wave energy is just sufficient) passing through samples of the complex are absorbed, causing d-electrons in lower-lying metal orbitals to be excited to higher-lying orbitals. Such electron movements are called transitions, and the d-block metals which show such behavior are commonly called transition metal elements.

15. Applications of coordination chemistry

Chemistry of complexes is encountered in qualitative analysis, sequestering of metal ions, and biological processes, among other applications.

16. Chelates

Chelation is the formation of five or six membered rings through the attachment of polydentate ligands to central atoms or ions. Chelates are much more stable than monodentate ligands. The increase in stability can be largely attributed to an increase in entropy during chelation. Addition of EDTA is a very effective means of sequestering metal ions. Sequestration of metal ions is one means of rendering metal ions less reactive. Chelation can also be used as a means of transporting heavy metals such as iron through soil as a plant food.

Quiz A

1. Ferromagnetism is a property:
   a. that permits certain materials to be made into permanent magnets
   b. that is exhibited by Fe, Co and Ni
   c. where the magnetic moments of the individual atoms are aligned in domains
   d. all of these
   e. none of these
2. What factor might favor +3 as a stable oxidation state of Fe?
3. Write a formula for dichromate ion.
4. What is the oxidation state of manganese in MnO_4^{2-}?
5. What is galvanized iron?
6. Name the following ion: $[CoCl_2(NH_3)_4]^+$.
7. What is the oxidation state of the central metal ion in $[FeF_6]^{3-}$?
8. Write the formula for hexachloroferrate(III) ion.

Quiz B

1. Are there unpaired electrons in the tetrahedral complex ion $[Co(CN)_4]^-$?
2. The complex $[Pt(NH_3)_2Cl_2]$ is a square planar complex. Are geometric isomers possible? Explain.
3. Define optical isomers.
4. The metal obtained from the ore rutile is:
   a. zinc

b. iron
c. titanium
d. mercury
e. copper.

5. What is the electron configuration of Fe^{3+}?
6. Write an equation for the reaction of Mn(s) with HCl(aq).
7. Complete and balance the following equation:

   Cr_2O_3 + Al ---> ?

8. Write the formula for sodium chromate.
9. What is the oxidation state of Mn in MnO_3^-?
10. Which element has the electron configuration $[Xe]4f^{14}5d^{10}6s^1$?

Quiz C

1. The d- and f- blocks of the periodic table make up:
   a. the inert metals
   b. the transition elements
   c. the active metals
   d. nonmetals
   e. none of these
2. Name three metals that form metal carbonyls.
3. Which transition metal has the electron configuration $[Ar]3d^64s^2$?
4. Why are Zn, Cd, and Hg not transition elements?
5. Write the electron configuration for V^{3+}.
6. Give two practical applications of coordination chemistry.
7. Why are many complex ions colored?
8. What is the oxidation state and coordination number of the central atom in $Ni(CO)_4$?
9. Define a coordinate covalent bond.

Self Test

1. Metal carbonyls are formed with CO and:
   a. Fe, Co and Ni
   b. Zn and Cd
   c. Cr, V
   d. all metals
2. Write the formula for chromium trioxide.
3. Elements in the lanthanide series have electron configurations which involve:

a. filling of the 4f subshell
b. filling of the 3d subshell
c. filling of the 4s subshell
d. filling of the 5d subshell

4. What is the coordination number of the central metal ion in $[CoCl_2(NH_3)_4]^+$?
5. Name the following ion: $[Cu(en)_2]^{2+}$.
6. Are the following two ions isomers? $[PtCl_2(NH_3)_4]^{2+}$ and $[Pt(NH_3)_4Cl_2]^{2+}$
7. What is the oxidation number of the central metal ion in the complex $[NiCl_4]^{2-}$?
8. Name the following ion: $[CoCl_2(NH_3)_4]^+$.
9. What are structural isomers?

CHAPTER 23

Chemistry and Life: More on Organic, Biological, and Medicinal Chemistry

This chapter looks at the nomenclature for organic compounds, reactions of different classes of organic compounds, some biological molecules and finally spectroscopy of biological molecules. This chapter serves as a good introduction for students who will be taking organic chemistry or biology courses.

Chapter Objectives: You will be able to:

1. Know how to name and write formulas from the names for alkanes, alkenes, alkynes, alcohols, and carboxylic acids.
2. Write typical combustion reactions for alkanes and substitution reactions for alkanes and alkenes.
3. Describe the reactions of alcohols.
4. Know the reactions of carboxylic acids.
5. Write the hydrolysis reactions of amides and esters.
6. Describe substitution reactions of benzene.
7. Know the substitution reactions of other aromatic compounds, including whether groups are ortho-, para-directing or meta-directing.
8. Describe the structure and general properties of lipids.
9. Describe carbohydrates and give some common examples.
10. Discuss proteins and know the general structure of amino acids and proteins.
11. Know the structure of amino acids and how they combine to form peptide bonds and proteins.
12. Know the basic structure and function of nucleic acids.
13. Describe IR, UV-VIS, and NMR spectroscopy.

Chapter Summary

1. Alkanes

Alkanes are compounds of carbon and hydrogen without multiple bonds. To name alkanes, we use one or more prefixes, a stem and an ending. The rules are given in the text and briefly summarized here:

1. The ending is *ane* for an alkane.
2. The stem name comes from the longest continuous chain (recall the names for stems given in Chapter 2).
3. Prefixes are used for groups attached to the stem or parent chain.
4. The stem or parent chain is numbered with the smallest Arabic numbers are used to indicate the position of the substituents.
5. If two or more identical groups are attached, use the prefixes di-, tri- etc.
6. Groups are listed in alphabetical order.

Example 23.1. Give the appropriate name for the compound

$CH_3CH(CH_3)(CH_2)_2CH\ (CH_3)CH_3$.

Solution: Writing out the structure gives $CH_3CHCH_2CH_2CHCH_3$ for the main chain, which is six carbons long. Using the rules above, we know that the ending will be *ane*, and the stem name from the six carbon chain is *hex*. The basic molecule is hexane, but we have to name and locate the substituents. There are methy (CH_3) groups on carbons 2 and 5. It does not matter which end of the chain we start numbering; the numbers for location of the substituents come out the same. Since the two groups are both methyl groups, the compound is 2,5 dimethylhexane.

2. Alkenes and Alkynes

To name alkenes and alkynes, note that there are a few simple changes to the rules for alkanes.

1. The ending for alkenes is *ene* and for alkynes is *yne*. Recall that alkenes contain a double bond and alkynes contain a triple bond.
2. The parent chain is the longest chain containing the multiple bond.
3. Where necessary we locate the position of the multiple bond by using a number.
4. Substituent groups are named as in alkanes and their position indicated by a number.

Example 23.2. Name the compound $CH_3CH_2CH{=}CHCH(CH_3)CH_2CH_3$.

Solution: This compound has a double bond and is an alkene. The ending is *ene*. The longest chain is seven carbon atoms, so the stem uses the prefix hept. The double bond is at carbon 3 so the compound in 3-heptene, but we need to name and locate the substituent. The CH_3 group is a methyl group and it is located at position 5. The compound is 5-methyl-3-heptene. Note that had we numbered the chain from the right side, the methyl group would have a lower number, but the location of the double bond would have a higher number. We use the number to locate the multiple bond with the lowest number.

3. Conjugated and Aromatic Compounds

A structure that has a series of alternate double and single bonds is said to be conjugated. Aromatic compounds have (4n + 2) pi-electrons in a conjugated, planar ring system. The most common aromatic compound is the six membered ring, benzene. Aromatic compounds have a conjugated bonding system that produces cyclic clouds of delocalized π electrons.

Exercise (text) 23.1B. Pyrrole is considered to be aromatic. Explain.

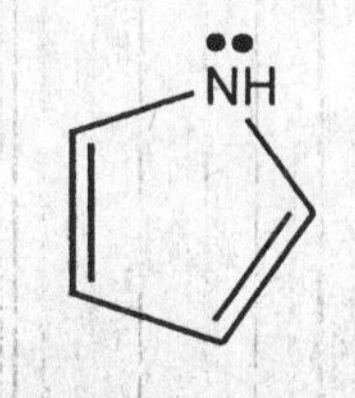

Pyrrole

Solution: The nitrogen in pyrrole has an unshared pair of electrons and can therefore complete the π electron cloud similar to a conjugated bonding system.

4. Alcohols

Alcohols are named in a manner corresponding to that described for alkenes and alkynes. The ending for an alcohol is *ol* and the alcohol group is given the lowest number when numbering the carbon stem.

Example 23.3. Name the following compound: $CH_3CH_2CH(CH_3)CH(CH_3)CH_2OH$.

Solution. The carbon stem is five carbons; therefore the prefix is pent. The compound is an alcohol, so the ending is *ol.* The stem compound is 1-pentanol, but we must name and locate the side groups. There are two methyl side groups, located at positions 2 and 3. The complete name is 2,3-dimethyl-1-pentanol.

5. Carboxylic acids

Carboxylic acids are named by the same procedures are described above with the ending of the alkane name changed by dropping the e and adding *oic acid.*

6. Reactions of aliphatic compounds

Alkanes undergo combustion reactions and substitution reactions. Alkenes are generally used as the starting materials for other organic products and undergo reactions to form polymers and substitution reaction.

Example 23.4. Write a reaction for the hydration of propene.

Solution: The reaction requires H_3O^+, which is usually supplied by sulfuric acid. Propene is an alkene with three carbon atoms. The reaction is

$$CH_3CH{=}CH_2 + H\text{-}OH \longrightarrow CH_3CH(OH)CH_3$$

In this case, the OH can add to the carbon atom and the product will be isopropanol or 2-propanol.

7. Reactions of alcohols

Substitution reactions are those in which another group substitutes for the alcohol. Halogens can substitute for OH under some conditions.

Alcohol groups also undergo dehydration in which the alcohol OH is removed and a multiple bond is formed with the removal of an additional H.

8. Reactions of Carboxylic Acids

Carboxylic acids react as acids in acid-base reactions and are involved in formation of esters and of amides.

Example 23.5. Write the equation for the reaction of ethanoic acid with methanol in the presence of H_2SO_4

Solution: This reaction will yield an ester.

$$CH_3COOH + CH_3OH \rightarrow \underset{\underset{\displaystyle O}{\|}}{CH_3COCH_3} + H_2O$$

9. Reactions of esters and amides

Esters and amides both undergo hydrolysis reactions. Esters are converted to an alcohol and carboxylic acid from which it was made. Amides are converted to the carboxylic acid and ammonia or the amine from which it was made.

10. Reactions of benzene

Benzene undergoes substitution reactions, rather than additions reactions typical of most unsaturated hydrocarbons. The mechanism for an electrophilic aromatic substitution is discussed in Section 23.7 of the text.

11. Reactions of other aromatic compounds

Substituted benzenes undergo reactions of the functional groups as do other aliphatic compounds. Electrophilic aromatic substitution is more complicated because, unlike benzene, all of the ring positions are not equivalent. The two positions adjacent to the substituent on the ring are ortho. The position directly opposite the substituent is the para position and the remaining two positions are the meta positions. Substituents are classified as to whether they will direct a substitution to the ortho, para positions or to the meta position. Some meta directors include NO_2, SO_3H, CHO, COR, COOH, and COOR.

Ortho and para directors include NH_2, NHR, NR_2, OH, OR, OCOR, R and X.

Example 23.6. Predict the major product(s) of the monochlorination of benzoic acid.

Solution: The carboxylic acid group is a meta director. The two meta positions are equivalent so the product will be 3-chlorobenzoic acid.

12. Lipids

The principal substances in living organisms are lipids, carbohydrates, proteins and nucleic acids. Lipids include fats and oils (triglycerides), cholesterol, some vitamins and components of cell membranes. Triglycerides are esters of fatty acids and glycerol. The esters of saturated fatty acids are usually fats and unsaturated fatty acids are oils. The hydrolysis of fats and oils in basic solutions yields soaps.

13. Carbohydrates

Carbohydrates are polyhydroxy aldehydes, polyhydroxy ketones or related compounds. Monosaccharides are the simplest carbohydrates and are usually found in ring form.

Polysaccarides, such as starch and cellulose, are polymers of monosaccharides. Optical activity is an important property of carbohydrates.

14. Proteins

Proteins are polymers of alpha amino acids. Amino acids are carboxylic acids with an amine group on the carbon next to the carboxyl group. The general structure is $RCH(NH_2)COOH$.

There are twenty different naturally occurring amino acids that serve as monomers in protein polymers. Amino acids are joined by peptide bonds to form polypeptides and proteins (Usually, longer aminoacid chains are proteins, shorter chains are referred to as peptides).
The peptide structure is given by the general formula

$$H_2N\text{-}CH(R_1)\text{-}CO\text{-}NH\text{-}CH(R_2)\text{-}CO\text{-}$$

where R refers to the side group of the amino acid. Most proteins fold into distinct three-dimensional shapes. The three-dimensional shape of a protein is referred to as the tertiary structure.

15. Nucleic Acids

Nucleic acids are polymers of nucleotides. A nucleotide consists of a sugar, ribose for RNA, deoxyribose for DNA, a phosphate group, and a cyclic amine base. The cyclic amine base is classified as either a purine or pyrimidine. The structure of DNA consists of two long chain polymers of nucleotides that are joined by hydrogen bonding into a double helix. DNA codes for a mRNA molecule, which determines the amino acid sequence of a protein.

16. Spectroscopy

Infrared spectroscopy (IR) detects the changes in vibrational modes of a molecule caused by the interaction of radiation and matter. Because vibrational modes are quantized, absorption of different frequencies of radiation can be used to determine the functional groups present in an organic compound.

Ultraviolet-visible spectroscopy detects pi-conjugated systems, including those in proteins and nucleic acids. Electrons in bonds are promoted to higher energy states.

NMR spectroscopy is based on the absorption of energy by nuclei of certain atoms in a molecule. If the atomic nucleus has an odd number for either the atomic number Z or the mass number A, the nucleus will be NMR active. These nuclei absorb electromagnetic radiation and change their orientation in a magnetic field. NMR spectrum pinpoints different "kinds" of H atoms for different locations and electronic environments in a molecule.

Quiz A

Fill in the blank

1. _______________ is the recognizable aromatic compound.
2. Compounds with five carbon chains have the prefix _______________.

3. A carboxylic acid functional group has the formula ______________.
4. ___________ substituted compound is an aromatic compound with substituents in a 1,2 relationship on the ring.
5. ______________ substituted compound is an aromatic compound with substituents in a 1,3 relationship on the ring.
6. The two functional groups always encountered in amino acids are ______________ and ______________.
7. The four different amine bases in DNA are ____________________.
8. Nucleotides consist of ___________, ___________ and _________________.
9. ______________________ spectroscopy measures the interaction of radiation and matter that causes changes in the vibrational modes of the molecules.
10. In order for a nuclei to be NMR active, it must have an ________ number protons and/or neutrons.

Quiz B

True or False

1. Lipids are defined by a distinct chemical structure.
2. Triglycerides are esters of glyerol and ribose.
3. Alkanes are hydrocarbons that contain at least one double bond.
4. Name alkynes by counting the longest chain of carbons that contain the double bond and add the suffix ene.
5. Alkanes react with oxygen during combustion to produce carbon dioxide and water.
6. A ketone contains a carbon-oxygen double bond.
7. Amines act as acids.
8. Carboxylic acids do not undergo ionization to act as acids.

Quiz C

Complete the following reactions:

1. $CH_2=CH_2 + Br_2 \rightarrow$
2. $CH_2=CH_2 + H_2O$ (in presence of H_2SO_4) $\rightarrow$
3. $CH_2=CH_2 + H_2 \rightarrow$
4. $CH_3CH_2CH_2OH + HI \rightarrow$
   (at high temperature)
5. $CH_3CH_2CH_2OH + H_2SO_4 \rightarrow$
   (at low temperature)
6. $CH_3CH_2CH_2OH$(excess) + $H_2SO_4 \rightarrow$
7. methyl,ethylester reacting with sodium hydroxide
8. monochlorination of toluene.
9. Benzoic acid reacting with HNO_3

Self Test

1. Write the structure for the following compounds:
   a. 2-nitrotoluene
   b. benzaldehyde
   c. 3-chloro-l-pentyne
   d. 2,3-dimethylbutanol
   e. diethylester
2. Describe the basic building blocks of proteins and give the general formula and bonding.
3. Which of the spectroscopic methods described would you expect to be able to determine the presence of an ester group in an organic molecule?
4. Draw the amino acid side chains for alanine, isoleucine, and serine.
5. If the sequence of base on one DNA strand is ATTCCG, what would be the sequence on the complimentary DNA?
6. What structural features of an organic molecule might cause it to give an absorption peak in the 200–400 nm range of the UV or visible region of the spectrum?
7. What features makes a nucleus suitable for NMR spectroscopy?

CHAPTER 24

Chemistry of Materials: Bronze Age to Space Age

This chapter first focuses on metals. The second part of this chapter describes two new types of materials: semiconductors and polymers.

Chapter Objectives: After completing this chapter you should be able to:

1. Define metallurgy and know the five principal types of ores.
2. Explain the processes for extraction of metals.
3. Describe pig iron and steel.
4. Know the principal ores of tin and lead and how they are processed.
5. Describe the processing of copper and zinc ores.
6. Discuss the processes involved in hydrometallurgy.
7. Explain the free-electron theory of metallic bonding.
8. Define band theory and use it to explain some of the properties of metals.
9. Use band theory to describe semiconductors and how they carry electricity.
10. Describe several natural polymers and their uses in other products.
11. Define addition polymerization and describe the steps involved in the reaction.
12. List the key features of condensation polymerization.
13. Describe a thermoplastic polymer, elastomer, and fiber.
14. Describe the use of biopolymers in medicine.

Chapter Summary

1. Metallurgy

Metallurgy is the general study of metals, including the extraction of metal from ores. The five principal types of ores and some examples are:

1. native ores (free metal): gold, silver and copper
2. oxides: iron, manganese, tin
3. sulfides: copper, nickel, zinc
4. carbonates: sodium, potassium, magnesium
5. chlorides (often as ions in solution): sodium, potassium, calcium

2. Extraction of metals

The processes described below are commonly required for extraction of metals:

1. Mining: The physical removal of the ore, usually either from deep mines for ores far below the surface or from pit mines, which are large and shallow pits for ores close to the surface.
2. Concentration: The ore is physically separated from the waste rock.
3. Roasting: Ores are heated to a high temperature and metal compounds are converted to metal oxides.

4. Reduction: This is an oxidation-reduction reaction where oxygen is removed from the metal oxide. Coke or coal is commonly used as a reducing agent. In most cases, the partial oxidation of carbon produces CO(g), which acts as the reducing agent.
5. Slag formation: High melting impurities are removed by formation of slag, a lower melting product. Slag is formed by basic oxides reacting with acidic oxides and amphoteric oxides.
6. Refining: The process of removing impurities by any of a number of chemical or physical means. Electrolysis is often used because it yields a pure product.

3. Pig iron and steel

Iron formed in a blast furnace is called pig iron. It is impure iron and most pig iron is used directly in the manufacture of steel. Solid metal obtained from pig iron is called cast iron. Pig iron and cast iron are impure and may be referred to as alloys. An alloy is a mixture of two or more metals or a metal with a nonmetal. Often the mixture is more desirable than the pure metal.

Most iron is converted to alloys known as steel. Pig iron is converted to steel by removal of carbon impurities to less than 1.5%. Other major impurities such as Si, Mn, P and S are removed as well as some minor impurities. Steel has more desirable properties than iron or pig iron.

4. Tin and lead

The tin ore is cassiterite, which is principally SnO_2. SnO_2 is then roasted with carbon to yield liquid Sn.

Lead is found as galena, PbS. The ore is roasted to yield PbO(s) and then reduced with coke. Several impurities are removed by taking advantage of the differences in the melting points of metals.

5. Copper and Zinc

Contamination of copper with iron is a significant problem. Copper ore is generally processed by partially roasting to convert iron sulfides to oxides. The roasted ore is heated to separate copper matte containing copper and iron sulfides in the bottom of the furnace and to remove slag containing oxides of iron, calcium, and aluminum. In a process called conversion, air is blown through the ore to convert any iron sulfide to iron oxide. The iron oxide is then removed.

Zinc occurs mainly as sphalerite (ZnS) and smithsonite ($ZnCO_3$). Both ores are converted to oxides by roasting.

6. Hydrometallurgy

Hydrometallurgy involves the processing of aqueous solutions of metal compounds and is generally less polluting than pyrometallurgical methods. Hydrometallurgy involves the following processes:

1. Leaching: Metal ions are extracted from an ore with a liquid.

2. Purification and concentration: Ion exchange and adsorption on activated charcoal are used to remove impurities. The solution is concentrated by evaporation.
3. Precipitation and reduction: Desired metals are precipitated as ionic solid. Metal ions are then reduced to free metal.

7. Free-Electron model of metal bonding

According to the free-electron model of metallic bonding, each atom in a metallic crystal loses its valence electrons, releasing them to the crystal as a whole. The metallic crystal is a lattice of positive ions with a "sea of electrons". The electrons are free to move throughout the crystal. The free-electron model explains many of the properties of metals, however it allows for specification of an electron's position and momentum more precisely than is permitted by the Heisenberg uncertainty principle. A quantum mechanical treatment is necessary.

8. Band Theory

The molecular orbital energy levels in a metal crystal are so small that the levels essentially merge into a band. This band is occupied by the valence electrons. A conduction band is a partially filled band of energy levels and is necessary for conduction of electricity. In some metals, the valence band is only partially occupied and is also the conduction band. In other cases, the filled valence band overlaps with an empty conduction band.

9. Semiconductors

If the energy gap between the valence band and the conduction band is large very few electrons can jump the gap and the material will be an insulator. In semiconductors, the energy gap is smaller and a significant number of electrons can jump the gap. The rate at which electrons jump the gap increases with temperature. Doping of a semiconductor increases the ability to conduct electricity. In an n-type semiconductor, the energy level of the electron donor is close to the conduction band and electrons are carried by the conduction band. In a p-type semiconductor, the energy level of the electron acceptor is close to the valence band and electric current is carried by positive holes in the valence band.

10. Natural polymers

Polysaccharides, proteins, and nucleic acids are all examples of natural polymers. Reaction of cellulose with sodium hydroxide and carbon disulfide leads to the production of rayon. The same reaction can also lead to the production of cellophane.

11. Addition polymerization

Addition polymerization is the process where monomers add to one another in such a way that the product contains all the atoms of the starting monomers. Furthermore, when the monomer is of low symmetry, such as for chloroethene, $Cl\text{-}CH{=}CH_2$, the monomers usually add in "head-to-tail" fashion, alternating the substituted and unsubstituted carbon atoms. The mechanism of addition polymerization can be characterized by the following steps:

1. Initiation. The reaction is started by an initiator, usually a free radical. This reaction leads to conversion of a double bond to a single bond with one unpaired electron. The product is a free radical.
2. Propagation. The radical formed in the initiator step joins another monomer to form a larger radical.
3. Termination. The propagation ends when a molecule is produced that does not contain a free radical.

12. Condensation polymerization

In condensation polymerization a small portion of the monomer is not incorporated in the final polymer. The key features of condensation polymerization include:

1. Each monomer contains at least two functional groups
2. The monomers are linked through the functional groups
3. Small molecules are formed as by products as the monomers are linked

Peptide bond formation is one example of condensation polymerization.

13. Physical properties of polymers

A thermoplastic polymer can be softened by heating and then formed into desired shapes. Physical properties are related to the structural features of the polymers, including whether they are linear polymers or branched chain polymers. Elastomers have a high degree of flexibility. A fiber is a polymer material obtained in a long, threadlike structure that generally has a high degree of tensile strength.

14. Biomedical polymers

Biomedical polymers have many applications including artificial joints, heart valves, replacement of arteries blocked by atherosclerosis, and treatment of burns.

15. Nanomaterials

A nanomaterial is any material that develops unique physical or chemical properties when the sample size is reduced to the nanometer scale. One of the applications of nanomaterials is as catalyists for reactions.

Quiz A

Fill in the blank.

1. ____________ is the science and technology of extracting metals from their ores.
2. Valence electrons in metals can be said to be ____________________, and not belong to any particular metal atom.
3. The number of molecular orbitals formed is ____________________ the number of atomic orbitals combined.

4. ______________ is a material that has an electrical conductivity intermediate between that of a metal and that of an insulator.
5. The separation of the valence band and the conduction band in terms of energy is called ______________________.
6. Conductivity of a metal ______________ with increasing temperature.
7. Three types of polymers are ________________, __________________, and ______________.
8. Two types of polymerization are ______________ polymerization and ______________ polymerization.
9. Proteins are examples of ______________________.

Quiz B

True or False

1. The free-electron model describes the electrical conductivity of polymers.
2. Tin ore is composed principally of SnO_2.
3. A doped semiconductor is one that has had a high insulating material added to it.
4. An insulator has a conduction band and valence band very close in energy.
5. The band gap is the difference in energy between the bonding and antibonding molecular orbitals.
6. Addition polymerization occurs when monomer units are combined such that part of the monomer is not included in the polymer.
7. Proteins are examples of polymers formed by condensation polymerization.
8. A fiber is a long threadlike polymer that has very low tensile strength.

Quiz C

1. What is the main ore of tin?
2. What is pig iron and how is it produced?
3. What is the principal compound of Zn in nature?
4. Write the reactions for the preparation of lead from the ore galena.
5. What is the main product of roasting? Write a reaction for the roasting of zinc carbonate.
6. Why must a band of energy levels be only partially filled to conduct electricity?
7. Define and give several examples of a polymer.

Self Test

1. Why must electrolytic reduction be used in the production of some metals such as lithium and sodium?
2. Use band theory to explain the electrical conduction of alkali metals.

3. Magnesium is a metal and conducts electricity, yet it appears that the valence band is completely filled. How do you explain this?
4. Write a reaction for the addition polymerization of ethylene.
5. Write a reaction for the condensation polymerization of two amino acids.
6. Describe n-type and p-type semiconductors.
7. How do high-density and low-density polyethylenes differ in structure and properties?
8. What is the chemical makeup of cellophane and how is it made?
9. Draw the structure of the "head-to-tail" polymer expected from $CH_2=CCl_2$.
10. Describe how termination occurs in addition polymerization.

CHAPTER 25

Environmental Chemistry

This chapter will look at the composition and properties of the Eacth's atmosphere and water supplies and some of the ways that human activities affect air and water.

Chapter Objectives: You will be able to:

1. Know the main components of the Earth's atmosphere.
2. Describe the layers of the atmosphere
3. Define relative humidity
4. Describe the process of nitrogen fixation
5. Describe the carbon cycle
6. Define air pollution and know the main source of the air pollutant, carbon monoxide.
7. Describe photochemical smog and be able to write equations for the reactions involved in removal of photochemical smog.
8. Describe industrial smog and contrast the difference between photochemical smog and industrial smog
9. Write the reactions for the production and destruction of ozone. Know the effects of the ozone layer on the Earth's atmosphere.
10. Describe the greenhouse effect.
11. List some of the major sources of the pollution of ground water.
12. Describe the problems of organic matter and sewage pollution in water.
13. Summarize the steps for wastewater treatment.
14. Define acid rain and describe some of the effects on the environment.
15. List some hazardous and toxic materials that are pollutants and describe their effects.

Chapter Summary

1. Composition of air

Air is composed mainly of nitrogen (78%) and oxygen (21%). Carbon dioxide and water vapor are minor components, but are vital for plants to produce food on which animal life depends.

2. Earth's atmosphere

Earth's atmosphere is divided into layers. The layer closest to Earth is the troposphere and contains about 90% of the mass of the atmosphere. It extends about 12 km above the surface. Weather and most human activity occurs in the troposphere.

The stratosphere is the next layer and extends about 12 to 55 km above the surface. The ozone layer lies in the stratosphere. The third and fourth layers are the mesophere and ionosphere, respectively. The ionosphere molecules absorb such high energy electromagnetic radiation that they dissociate into atoms and positive and negative ions.

3. Water vapor

Water vapor is present in the air. We frequently speak of the water vapor as the relative humidity. The relative humidity is the partial pressure of water vapor/vapor pressure of water x 100%.

4. Nitrogen fixation

Nitrogen from the atmosphere must be converted to compounds that are more readily usable by living organisms. The conversion is know as nitrogen fixation. The process of nitrogen fixation occurs by bacteria converting atmospheric nitrogen to ammonia. Plants take up the nitrogen in the form of nitrate or ammonium ions. Nitrogen in plants combines with carbon compounds in photosynthesis to form amino acids. Decay of plant life returns nitrogen to the soil as nitrates and ammonia.

5. The carbon cycle

Photosynthesis converts CO_2 to carbohydrates. Animals acquire carbon compounds by consuming plants or other animals. CO_2 is returned to the atmosphere by respiration.

6. Carbon monoxide

Air pollution results when a substance in found in air in greater abundance than it occurs naturally and the substance has harmful effects on human health or the environment. Carbon monoxide is an air pollutant and results mainly from the incomplete combustion of hydrocarbons in gasoline. Carbon monoxide exerts its effects by replacing oxygen bound to hemoglobin, the red blood pigment.

7. Photochemical smog

Photochemical smog is produced by the interaction of sunlight with air the contains nitrogen oxides, hydrocarbons and other pollutants. Photochemical smog has a higher than normal concentration of ozone. Smog starts with the production of NO, often from burning of fossil fuels. NO is then oxidized to NO_2 which decomposes. Oxygen atoms from the decomposition react with oxygen molecules to form ozone, O_3. NO and NO_2 react with hydrocarbons to produce other harmful compounds. The NO content of automobile exhaust can be lowered by catalytic converters which reduce NO to N_2. Reduction can also be accomplished by using a mixture of fuel and air which produces CO that reduces NO to N_2 and CO_2.

8. Industrial smog

Industrial smog is usually produced by industrial processes and contains high levels of sulfur oxides. Industrial smog can be removed by collector plates where smokestack gases are given an electric charge and removed on an oppositely charged plate. Removal of sulfur containing minerals before burning will also reduce industrial smog.

9. Ozone layer

The ozone layer is a band of the stratosphere that has a much higher ozone content than the rest of the atmosphere. Ozone protects the Earth from the harmful effects of UV radiation. Ozone is formed by the reaction of oxygen with UV radiation. Ozone can be destroyed by the reaction with NO or with chlorofluorocarbons.

10. Greenhouse effect

The greenhouse effect is the process where the Earth's atmosphere is warmed by radiant energy that is reflected back from the Earth and absorbed by substances in the atmosphere, much as a greenhouse is warmed. Some warming is necessary to maintain the temperature on Earth, however, increases in carbon dioxide could enhance the greenhouse effect and cause changes in the global climate.

11. Natural waters

The Earth's surface is largely covered by water. The hydrosphere is the water on the Earth's surface including oceans, lakes, and streams. Most of the water is salt water and cannot be used for drinking or industrial purposes. Pollution of fresh water has become a problem with increasing population and industry. Until about a hundred years ago, biological pollution by microorganisms from human wastes was a severe problem. Cholera and typhoid were serious problems. These diseases and others are still a problem in developing countries.

Another major source of pollution is from fertilizers and pesticides used in agriculture and on lawns. Chemicals buried in dumps are also a source of pollution. A particular problem is thrichloroethene which is often used as a dry cleaning solvent. Even though it is soluble in only ppm range, it is a suspected carcinogen and even trace amounts are a problem. Leakage of underground storage tanks for gasoline and spillage of oil from tankers are other sources of pollution.

12. Chemistry and biology of sewage

Disease from dumping of sewage into waterways is not the only problem. The organic matter in sewage is decomposed by bacteria and depletes the oxygen. Organic matter can be degraded either by aerobic or anaerobic. Aerobic oxidation takes place in the presence of oxygen and can result in depletion of oxygen. If enough oxygen is depleted, fish and other life forms can no longer survive in the water. The biochemical oxygen demand (BOD) measures the quantity of oxygen needed for the oxidation of organic compounds. Even when there is enough oxygen, oxidation products such as nitrates can serve a nutrients for algae which die, become organic waste, and increase the BOD through eutrophication.

13. Water Treatment

The process of water treatment usually occurs in these steps:

1. Treatment with slaked lime and a flocculating agent to form a gelatinous precipitate that removes dirt and bacteria
2. Filtering through sand or gravel
3. Aeration removes odorous compounds and improves the taste of water
4. Chlorination to kill remaining bacteria. Fluorides may be added to prevent tooth decay.

These processes are often carried out at wastewater treatment plants. A summary of wastewater treatment methods is given in Table 25.6.

14. Acid Rain

Acid rain is the result of sulfur oxides that are converted to sulfuric acid and nitrogen oxides that are converted to nitric acid in the atmosphere. Acid rain is any rainfall that is more acid than it would be from CO_2 dissolved in the atmosphere. Acid rain has many detrimental effects including corrosion, effects on life in lakes and streams, and crop and forest yields.

15. Poisons and hazardous materials

Almost anything can be a poison if the dose is large enough. Poisons and hazardous materials can be classified based on their properties: corrosivity, ignitability, reactivity and toxicity. Some examples of hazardous materials are:

1. ozone: a strong oxidizing agent
2. acids and bases: corrosive
3. carbon monoxide: a health hazard because it replaces oxygen in the blood
4. cyanide: a poison because it shuts down cell respiration
5. heavy metals: often interfere with enzyme reactions
6. carcinogens: slow poisons that trigger malignant cell growth

Quiz A

True/False

1. The most abundant component of the atmosphere is CO_2.
2. The layer of the atmosphere nearest the Earth's crust is the troposphere.
3. Weather occurs in the stratosphere.
4. The greenhouse effect refers to the increase in plant growth on the Earth's surface.
5. Nitrogen cannot be used directly by higher plants and animals.
6. Carbon monoxide is one form of air pollution.
7. Sewage waste is a problem because is contains organic material that decays.
8. Photochemical smog has a higher than normal concentration of molecular oxygen.
9. Chlorine is added in water treatment to prevent tooth decay.

10. Aerobic oxidation occurs in the presence of dissolved oxygen.

Quiz B

Fill in the blank

1. ____________________________ measures the quantity of oxygen needed for the oxidation of the organic compounds in 1 liter of water.
2. Most organic mater is degraded by ___________________________.
3. In waste water treatment, ____________________ is used to kill any remaining bacteria.
4. Acid rain results from ____________ and _________________ which are converted to acids in the atmosphere.
5. Ozone in the stratosphere protects living organisms by ________________________.
6. The ozone layer is threatened by the release of ______________________ into the atmosphere.
7. ______________ is the study of the response of living organisms to poisons.
8. Heavy metal ions are toxic because they _______________________.
9. Carbon monoxide is poisonous because it interferes with the transport of ___________________.
10. Relative humidity is equal to ____________________________________ x 100%.

Self Test

Matching

1. acid rain	a. the condensation of water vapor on a solid followed by solution formation
2. carbon cycle	b. the water vapor content of air
3. aerobic oxidation	c. oxidation in the presence of oxygen
4. global warming	d. rainfall that is more acidic than is water in equilibrium with atmospheric carbon dioxide
5. toxicology	e. the sum of all the processes by which carbon atoms are cycled through out Earth's solid crust, oceans, and atmosphere
6. deliquescence	f. the anticipated increase in Earth's average temperature resulting from the increase in carbon dioxide and other infrared-absorbing gases into the atmosphere
7. humidity	g. the ability of carbon dioxide and other gases to absorb and trap energy radiated by Earth's surface as infrared radiation.
8. greenhouse effect	h. the study of the effects of poisons, their identification or detection, and the development of antidotes.

Short Answer

9. Describe the layers of the atmosphere including which layer contains the ozone layer, where weather occurs, and the layer closest to the Earth.
10. Describe the greenhouse effect and some of the consequences of an increase in the Earth's temperature.
11. Outline the steps for treatment of waste water.
12. Which compounds are important contributors to photochemical smog?
13. Describe the toxic effects of CO, heavy metals, and cyanide.

Answers to Quizzes

Chapter 1

Quiz A

1. 1 kg = 10^3 g/kg x 10^6 μg/g = 10^9 μg
2. four significant figures
3. 4.0 in. x 2.54 cm/in. x 6.0 in. x 2.54 cm/in. = 150 cm^2 (two significant figures)
4. (a) element
5. 2.54 g x cm^3/1.354 g = 1.88 cm^3
6. (a) 3 °C
7. (d) heterogeneous mixture
8. (c) can be tested
9. 22 °C, 295 K

Quiz B

1. (a) 6, (b) 2, (c) 3
2. (b) carbon dioxide freezes to form dry ice, solid carbon dioxide.
3. (c) a law
4. (a) 298.0 K, (b) 86 °F
5. (a) 1.23 μg, (b) 1.230 L, (c) 4.78 Mm
6. (a) 1.07 m, (b) 0.412 oz, (c) 0.0154 lb

Quiz C

1. False. The more precise the measurement will be.
2. False. 15 °C is 59 °F.
3. True
4. True
5. 488.2
6. 10 lbs x 453.6 g/lb x 1 cm^3/19.8 g = 229 cm^3
7. 0.010 L
8. 10000 m x 1Km/1000 m x 0.6214 mi/Km x 4.0 min/mi = 25 min
9. 0.0001 mg
10. 194 °F, 363 K

Self Test

1. a. chemical b. physical c. physical d. chemical e. physical

2. a. heterogeneous
   b. homogeneous
   c. heterogeneous
   d. pure substance
   e. pure substance
   f. pure substance
3. a. 102,000 mm b. 450,000 mL c. 90 cm d. 0.015 kJ
   e. 15 m/s x 1 Km/1000 m x 0.6214 mi/Km x 60 s/min x 60 min/hr = 34 m/hr
4. d.
5. c.
6. b.
7. b.
8. hypothesis –c
   accuracy – f
   density – e
   element – b
   precision – g
   volume – a
   theory – d
9. 7 ft x 12 in/ft x 1m/39.37 in = 2.1 m
10. 500 mL x 19.3g/mL = 9650 g x 1 mL/g = 9650 mL or 9.65 L
11. 530 cal/hr x 1hr/60min x 25 min = 221 cal, 1 hotdog/160 cal x 221 cal = 1.4 hotdogs
12. 173^0F
13. t = 9/5 t + 32, solve for t. t = -40
14. 1 gal x 3.785L/gal x 1000 mL/L x 1 cm^3/mL x 1 g/cm^3 = 3785 g
15. 500 mi x 1 gal/35 mi = 14.3 gal, 14.3 gal x \$2.00/gal = \$28.60
16. 10 ^{0}C, C degree is larger than a degree F.
17. 1.80 g/L

Chapter 2

Quiz A

1. (a) proton
2. (c) 17 protons and 18 neutrons
3. a. potassium cyanide b. calcium carbonate c. sodium sulfate
4. a. HCl b. H_2SO_4 c. NaOH d. H_3PO_4
5. (c) conservation of mass

6. (a) same number of protons
7. a. $OH\text{-}CH_2\text{-}CH_2\text{-}CH_2\text{-}CH_2\text{-}CH_3$

   b. $CH_3\text{-}CH\text{-}CH_2\text{-}CH_2\text{-}CH_2\text{-}CH_3$ (with CH_3 attached to the CH)

   c. $H_3C\text{-}CH_2\text{-}CH_2\text{-}COOH$

Quiz B

1. (d)
2. (d) isotopes
3. 1-methylpropane
4. a. sodium hydroxide b. lithium sulfate c. hydrochloric acid d. methanol
5. a. $FeCl_3$ b. SF_6 c. H_2SO_4 d. Na_3PO_4
6. (b) multiple proportions
7. $^{35}_{17}Cl$

Quiz C

1. d.
2. d.
3. f.
4. (a) aluminum phosphite

   (b) cesium iodide

   (c) cobalt(III) carbonate

   (d) hydrosulfuric acid
5. d.
6. (a) CH_3CH_2OH

   (b) C_6H_{14}

   (c) $CH_3CH_2CH_2O$

   (d) H_3CNH_2

   (e) H_3CCH_2COOH
7. $(0.5183)(106.905) + (0.4817)(108.905) = 107.868 = 107.9$
8. Acids: simple acids produce H^+ in aqueous solution

   Bases: simple bases produce OH^- in aqueous solution
9. Since the atomic mass is much closer to 300 than 400, the 300 amu isotope must be more abundant. If they were equally abundant, the atomic mass would be 350 amu.
10. Ionic: NaCl, KCl, $Ca(OH)_2$, etc.; molecular: CH_4, CCl_4, CH_3OH. See text for more examples.

Self Test

1. (a) $Fe(IO_4)_3$
   (b) CaF_2
   (c) CH_3CH_2OH
   (d) HBr
   (e) K_3PO_3
2. c
3. b
4. d
5. b
6. False, same atomic number, different mass number
7. True
8. False. Different number of protons
9. True
10. False, correct formula is $HBrO_2$.
11. No. Law of definite proportions.
12. $(0.7577)(35) + (0.2423)X = 35.45$
    $X \approx 37$
13. ZnS is approximately 67% Zn by mass, therefore (0.67)(0.308) = 0.206 g Zn in ZnS produced. And 0.294 g remain.
14. 207.177
15. Approximate % is 50% mass protons (6), 50% mass neutrons (6) and << 1% mass electrons which have < 1/1000 mass of proton or neutron.
16. There is only one ion of calcium with a +2 charge. Iron can have +2 or +3 ions so we must specify which is present.
17. 

18. litmus paper, reaction with metals, etc.
19. Metals are on the left of the periodic table. Properties include: malleable, good conductors, tend to lose one or more electrons in forming compounds.

Chapter 3

Quiz A

1. 40.0% C, 6.7% H, 53.3% O
2. 5.72 g

3. 53.5 g/mole
4. $C_3H_8O_2$
5. $2\ NH_3 + 5/2\ O_2 \text{ --> } 2\ NO + 3\ H_2O$
6. 3.94 g
7. 0.063 M
8. 0.38 g
9. 840 g

Quiz B

1. 324.6 g/mole
2. N_2O
3. 1.67×10^{23} molecules
4. 0.829 L
5. $AlCl_3 + 3\ NaOH \text{ --> } Al(OH)_3 + 3\ NaCl$
6. 91.7%
7. 3.75 M
8. $4\ BF_3 + 3\ H_2O \text{ --> } H_3BO_3 + 3\ HBF_4$
9. 0.08 g

Quiz C

1. True
2. False, the molecular formula may be a multiple of the empirical formula.
3. True
4. False, gives percent composition.
5. False, smaller than theoretical yield.
6. $CH_3CH_3 + 3\ 1/2\ O_2 \rightarrow 3H_2O + 2CO_2$
7. CaF_2, 40.078 + 2 (18.9984) = 78.075
8. 0.0103 M
9. 23 g Na x 1 mol/23.00g = 1 mol Na Similar calculations for the other elements gives 1 mol Na, 1 mol H, 1 mol C and 3 mol O. Formula is $NaHCO_3$. Cannot tell the molecular formula from the information given.
10. (137.327) x 3 + 8 (16) + 2 (31) = 602 g/mol
    9.0 g x 1 mol/602 g = 0.015 mol

Self Test

1. mole- e
2. molarity – a

3. theoretical yield – h
4. chemical equation – g
5. empirical formula – c
6. combustion reaction – b
7. Avogadro's number –f
8. microscopic level – j
9. solvent - i
10. molar mass - d
11. $PCl_3 + 3H_2O \rightarrow H_3PO_3 + 3HCl$

    0.89 g PCl_3 x 1 mol/137.33 g = 0.0065 mol

    For each mole of reactant, one mole of product is produced. Therefore 0.0065 mol H_3PO_3 is produced and 0.53 g . For a 65% yield 0.35 g will be formed.
12. 0.50 mol/L x 0.250 L x 40.0g/mol = 5.0g
13. $2AgNO_3 + Na_2SO_4 \rightarrow Ag_2SO_4 + 2NaNO_3$
14. 0.507g C x 1mol/12.011 g = 0.0422 mol C

    0.0425g H x 1mol/1.00794 g = 0.0422 mol H

    0.451g O x 1 mol/15.9994g = 0.0282 mol O

    Dividing all values by 0.0282 gives

    $C_{1.5}H_{1.5}O$ and multiplying by 2 to get whole numbers, $C_3H_3O_2$ for the empirical formula. The empirical formula weight is 71 so the molecular formula is 2 x the empirical formula, $C_6H_6O_4$.
15. 12M HCl x V = 0.25M x 0.1L

    V = 0.002 L or 2.0 mL
16. 3.45 mol x 40g/mol = 138 g
17. mass of CH_3OH in 1.0 L of 1.5 M CH_3OH
18. $CaCN_2 + 3H_2O \rightarrow CaCO_3 + 2NH_3$

    0.0399 mol produced
19. $CaCO_3$ is 40% Ca by mass. Therefore if we need 0.200g Ca, 0.200 = 0.40 X and X = 0.500g. If the antacid is only 75% calcium carbonate, then 0.75 X = 0.500 and X = 0.667g antacid.

    0.200 g Ca x 1g $CaCO_3$/0.40 g Ca x 1 g antacid/0.75 g $CaCO_3$ = 0.667 g antacid
20. H = 8/(8 + 12 x4) x 100% = 14.3%

    C = 48/56 x 100% = 85.7%

Chapter 4

Quiz A

1. c. HCl

2. c. 0.005 M $AlCl_3$
3. oxidation number increases
4. +3
5. no reaction
6. A strong acid is fully ionized while a weak acid is only partially ionized. An example of a strong acid is HCl and an example of a weak acid is acetic acid, CH_3COOH.
7. 46 mL
8. A nonelectrolyte is a substance that is not ionized and does not conduct electricity. An example is CH_4.
9. one ionizable H atom
10. $NH_3 + H_2O \rightarrow NH_4^+ + OH^-$

Quiz B

1. $2\ BrO^- + 2\ H_2O + 2\ e^- \rightarrow Br_2 + 4\ OH^-$ reduction occurs
2. oxidation number increases
3. +4
4. $\rightarrow BaSO_4 + CuS$
5. d. 0.008 M $Ca(NO_3)_2$
6. c. CH_3COOH
7. 3.33 mL
8. $Mg^{2+} + 2\ OH^- \rightarrow Mg(OH)_2(s)$
9. O.S. = +2
10. Oxygen = +2, fluorine = -1.

Quiz C

1. d
2. a
3. d
4. b
5. True
6. True
7. False
8. False
9. True
10. True

Self Test

1. a. strong b. strong c. weak d. nonelectrolyte e. nonelectrolyte
2. a. insolubleb. soluble c. insoluble d. insoluble
3. $Ca(OH)_2 + CH_3COOH \rightarrow$
   $Ca^{2+} + 2OH^- + 2CH_3COO^- + 2H^+ \rightarrow 2H_2O + Ca^{2+} + 2CH_3COO^-$
   Net ionic: $2OH^- + 2H^+ \rightarrow 2H_2O$
4. $Ca(OH)_2 + 2HNO_3 \rightarrow Ca(NO_3)_2 + 2\ H_2O$
5. 0.025 L x 0.05mol/L x 1L/0.214 mol = 0.00584 L or 5.84 mL
6. 0.1mol/L x 0.010L x 2 (OH^-)/mol x 1L/6 mol = 0.00033L = 0.33 mL
7. a. no ppt b. $Ca^{+2} + S^{-2} \rightarrow CaS$ c. $Mg^{2+} + CO_3^{2-} \rightarrow MgCO_3$
8. a. N, -3; H, +1 b. C, -4; H, +1 c. S, +6; O, -2 d. H, +1; O, -1
9. $2Fe_2O_3 + 3C \rightarrow 4Fe + 3CO_2$
10. Yes, $Fe^{3+} \rightarrow Fe$ reduction, gain of e^-; $C \rightarrow C^{4+}$ oxidation, loss of electrons
11. $Cu^{2+} + Zn \rightarrow Zn^{2+} + Cu$
12. a. c.
13. d.
14. no, organic compounds do not release H^+
15. reducing agent is oxidized
16. no. If vitamin E prevents oxidation, it is probably oxidized itself, and therefore a reducing agent.

Chapter 5

Quiz A

1. False
2. True
3. False
4. 1440 mmHg = 1440 torr
5. 2.2 g
6. 8.6 L
7. pascal
8. Gases have a lot of empty space between molecules
9. Actual volume is larger because gas molecules occupy space.
10. P increases with T.

Quiz B

1. (d.) all gases the same
2. (a.) H_2
3. (c.) Ar

4. 83.8 g/mole
5. 17.9 g
6. Rate O_2: Rate Xe = 2.03
7. n = PV/RT, n = 0.0450 mol
8. 273.15 K, 1 atm
9. $V \propto T$

Quiz C

1. pressure – f
   Boyle's Law – b
   Ideal Gas Law – c
   Partial pressure – g
   Kinetic Molecular Theory – h
   Diffusion – d
   Effusion – e
   Charles' Law – a
2. P = g x d x h
   Setting the equation for water equal to that for Hg and solve for the height of Hg.
   g/mL x 2.8 m = 13.6 g/mL x h_{Hg}
   h_{Hg} = 205.9 mm, 205.9 mm Hg x 1 atm/760 mm Hg = 0.271 atm
3. Since V and T are directly related, when V increases by two, T must increase by two. T = 40 ^{0}C.
4. CH_4 is lighter and therefore diffuses faster.
5. 34.2 L

Self Test

1. False, V is directly proportional
2. False, P is inversely proportional to V.
3. True
4. False, lighter gas diffuses faster.
5. False, Pascal
6. False, 273.15
7. True
8. 0.75 atm, 0.15 atm, 0.10 atm
9. $$\frac{\mu_H}{\mu_{Cl}} = \sqrt{\frac{M_{Cl_2}}{M_{H_2}}}$$

   μ_{Cl} = 309.9 m/s

10. High pressure, low temperature
11. 0.780 mol x 22.4L/mol = 17.5 L
12. T = MP/dR = 331 K
13. a. V increases

    b. V increases

    c. V increases
14. $P \propto 1/V$ As P decreases, V increases. Molecules are in random motion, so as force compressing molecules is decreased, they will move about to fill the space and collide with the walls of the container.
15. Stoichiometry for the reactant and product is 1:1. So $P_1V_1/RT_1 = P_2V_2/RT_2$ The pressure doesn't change, so canselling P and R, and solving for V_2

    50 L x 873K/373K = 117 L
16. The ratio of the rates will be the inverse ratios of the square roots of the mass. $r_{Ar}/r_{Ne} = 0.711$

Chapter 6

Quiz A

1. (c)
2. 94 °C
3. $H_2(g) + 1/2\ O_2(g) \dashrightarrow H_2O(l)$
4. 3.41 J/g °C
5. -2.8 kJ/mol
6. (a) quantity of heat
7. (d) all of the above
8. +157.3 kJ

Quiz B

1. +87.9 kJ/mol
2. 28 kJ
3. 5.33 kJ
4. A reaction in which heat is given off to the surroundings.
5. Internal energy decreases by 180 J.
6. C(graphite) + 2 $H_2(g)$ --> $CH_4(g)$
7. (a) energy transferred as the result of a temperature difference
8. Energy increases by +175 J.
9. calorimeter

Quiz C

1. heat
2. potential energy
3. law of conservation of energy
4. positive
5. energy
6. $\Sigma\Delta H_f$ (products) - $\Sigma\Delta H_f$(reactants)
   = -814.0 kJ/mol –(-395.7 kJ/mol +(-285.8 kJ/mol)) = -132.5 kJ/mol
7. $C = q/\Delta T = 1200\ J/30\ ^0C = 40\ J/^0C$
8. Internal energy decreases by 50 J
9. $3C(gr) + 3H_2(g) + 1/2\ O_2(g) \rightarrow C_3H_5OH(l)$

Self Test

1. f.
2. e.
3. c.
4. a.
5. b.
6. d.
7. (-184.6 kJ/2) = -92.3 kJ mol^{-1}
8. 24.9g x 1mol/2g = 12.45mol
   12.45 mol x –571.6 kJ/2mol = -3558 kJ
9. –571.6 kJ/mol O_2 x Xmol = 500 kJ
   X = 0.875 mol
10. 16.7 J
11. $N_2H_4 + 2O_2 \rightarrow 2H_2O_2 + N_2$
    Numbering equations as 1,2, and 3 to calculate the enthalpy, use (1) + 2x (3) – 2 x(2) to get
    -425.6 kJ

Chapter 7

Quiz A

1. (c) ultraviolet radiation
2. 4.85×10^{-12} m = 4.85 pm
3. True
4. False, loses energy
5. True

6. 1.937×10^{-18} J
7. $l = 0$
8. 6.0×10^{-8} m
9. 9.5×10^{-21} J

Quiz B

1. 122 nm
2. (a) possible
   (b) not possible, s can only have values of +1/2 and -1/2
   (c) not possible, l cannot be equal to n.
   (d) possible
3. 10 electrons
4. 3.68×10^{-19} J
5. 1.06×10^{7} m/s
6. (b) frequency
7. $3.0 \times 10^{8}\ ms^{-1}$
8. 3 orbitals
9. $l = 2$, d subshell has 5 orbitals.

Quiz C

1. electrons
2. Thomson
3. Millikan
4. Mass spectrometry
5. Nucleus or neutrons and protons
6. Frequency
7. Bohr's atomic theory
8. $E = hc/\lambda = 3.75 \times 10^{-19}$ J
9. $l \leq n - 1$, for f subshell, $l = 3$, therefore n = 4
10. n = 4, $l = 1$; $m_l = -1,0,1$; $m_s = \pm 1/2$

Self Test

1. $m_l = -2$ to $+2$
2. $\Delta E = -B/n^2$, n = 1, $\Delta E = -2.179 \times 10^{-18}$ J
3. $\nu = 1.28 \times 10^{13}$ Hz $= c/\lambda$

   $\lambda = c/\nu = (3.00 \times 10^{8} m\ s^{-1})/(1.28 \times 10^{13}\ s^{-1}) = 2.34 \times 10^{-5}$ m = 2340 nm
4. $(200$ mi$/3.0 \times 10^{8}$ m $s^{-1}) \times 1$ km/0.6214 mi $\times 10^{3}$ m/km $= 107.2 \times 10^{-5}$ s = 1.07 ms

5. $\Delta E = 3.03 \times 10^{-19}$ J = hc/λ λ = 656 nm
6. $m = h/\lambda\mu = 6.63 \times 10^{-34}$ kg m^{2}s^{-2} /(0.265 x 10^{-9} m x 2.74 x 10^6 ms^{-1}) = 9.13 x 10^{-31} kg
7. true
8. true
9. false, speed of light is constant
10. false, inversely
11. false, energy level
12. false, energy is absorbed to go to higher level
13. electron spin
14. quantized
15. wavelike, particle-like
16. line spectrum
17. inversely
18. Bohr

    a) Electron moves in circular orbit around nucleus, only certain orbits are allowed

    b) Energy of electron increases further it is from the nucleus.
19. Black-body radiation is electromagnetic radiation emitted by solids. Frequency increases as temperature increases.
20. Emission of an electron from the surface of certain materials when struck by light of the appropriate frequency.

Chapter 8

Quiz A

1. $1s^22s^22p^63s^2$
2. b. Na
3. Na > Mg
4. $F^- > Na^+$
5. Cs < Rb < Na
6. lose one
7. Nitrogen
8. ns^2np^6
9. $1s^22s^22p^63s^23p^6$

Quiz B

1. $1s^22s^22p^63s^23p^63d^54s^1$
2. Na > Mg > Al
3. Ar, no unpaired electrons.

4. Ca
5. Ga < Se < S.
6. Br^-
7. (b) a vertical column
8. No two electrons in the same atom can have all four quantum numbers the same.
9. Ne
10. five

Quiz C

1. 2, $[Ne]3s^23p^2$
2. Rb, metallic character decreases left to right.
3. Xe, radii increase top to bottom
4. True
5. False, Hund's rule states that orbitals fill singly before pairing of electrons.
6. False, singly
7. True
8. False, n and *l*
9. False, highest energy
10. True
11. False, 1A and 2A

Self Test

1. b.
2. c.
3. a.
4. d.
5. h.
6. j.
7. i.
8. e.
9. f.
10. g.
11. $[Ne]3s^23p^5$, seven
12. ns^2np^3 p-block
13. electron is one shell further from nucleus, so there is less attraction
14. second, after one electron is removed there is a greater attraction to nucleus because of the positive charge. The first electron is removed from a neutral atom and the second from a positively charged atom.

15. The effective nuclear charge increases and valence electrons are pulled toward the nucleus.
16. ns^2np^6
17. small number of electrons in valence shell
18. decrease, The greater the distance between the atomic nucleus and the electron, the less tightly it is held.

Chapter 9

Quiz A

1. $H-\overset{\bullet\bullet}{\underset{\bullet\bullet}{O}}-\overset{\bullet\bullet}{\underset{\bullet\bullet}{Cl}}\,:$
2. (a) N_2
3. (b) ICl
4. -420 kJ
5. $\overset{\bullet\bullet}{\underset{\bullet\bullet}{O}} = C = \overset{\bullet\bullet}{\underset{\bullet\bullet}{O}}$
6. P< O < F
7. A covalent bond is formed when electrons are shared between atoms.
8. Ar has an outer shell octet of electrons which is a stable configuration. It therefore does not share, gain, or lose electrons to form bonds.
9. H-C= C-H

Quiz B

1. (c) $CaBr_2$
2. (b) resonance
3. H-C=C-H
     | |
    H H
4. B < N < F
5. This would be a resonance between a triple and single (two structures) and two double bonds. The bond lengths would be close to double bond lengths.
6. $H_2C=CH_2$

7.

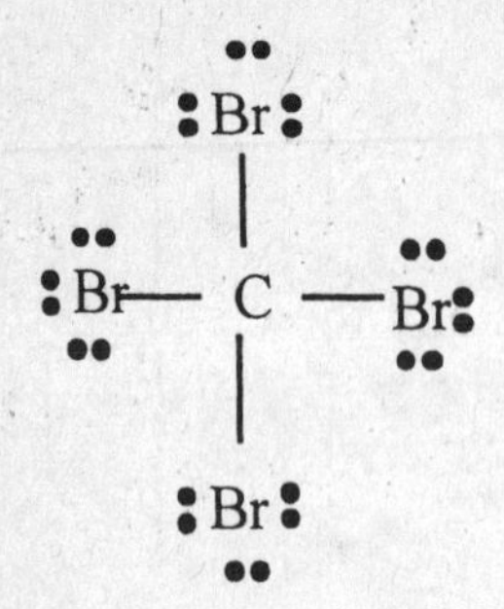

8. [O:H]$^-$
9. The transfer of electrons between metal and nonmetal atoms results in an ionic bond (as in NaCl).
10. An expanded octet describes the situation in which certain atoms can accommodate more than eight electrons in their valence shells when forming bonds. An example is PCl_5.

Quiz C

1. bond energy
2. Lewis structure
3. Electronegativity
4. Ionic bonds
5. Covalent bond
6. False, unshared pair
7. True
8. False, number of pairs of electrons shared in a bond
9. False, shorter bonds are stronger
10. True

Self Test

1. e
2. h
3. g
4. d
5. b
6. c
7. a
8. f

9.

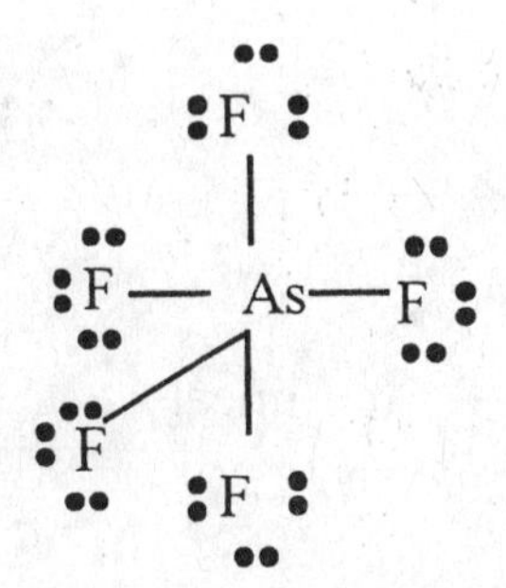

10.

```
    F              F
    |              |
    B- F           B=F
    |              |
    F              F
```

(1) has no formal charge and therefore is preferred.

11. b. B-N has greatest difference in electronegativity

12. Cl^-, :Cl: Na^+ has [Ne]; $1s^2 2s^2 2p^6$

13. $CH_4 + O_2 \rightarrow CO_2 + 2H_2O$ bonds broken = 4 x C-H + 2x O=O; bonds formed 2 x C=O + 4x H-O; $\Delta H = \Sigma BE(\text{reactants}) - \Sigma BE(\text{products}) = -676$ kJ

14.

O= S — O ⟷ O—S = O

15. single < double < triple < ionic bonds

Chapter 10

Quiz A

1. a. linear b. bent c. linear
2. trigonal bipyramid electron geometry, T shaped molecular geometry, sp^3d hybridization
3. (d) hybridization
4. pi bonds involve side-by-side overlap of p orbitals.
5. (d) has both sigma and pi bonds

6. The bond order is one-half the difference between the number of electrons in bonding MOs and number in antibonding MOs.
7. two
8. Valence bond method describes a covalent bond as the overlap of atomic orbitals of the bonded atoms in a region between the atomic nuclei.
9. Four hybrid orbitals will be formed.

Quiz B

1. (a) C_2H_4
2. (a) trigonal bipyramid (b) bent, <109 (c) linear
3. (c) BCl_3
4. (b) strong bonds
5. sp^3 hybridization, tetrahedral arrangement.
6. The combination will be sp^3d.
7. Antibonding orbitals are higher energy than bonding orbitals.
8. Two molecular orbitals result.

Quiz C

1. True
2. False, d-block elements
3. True
4. True
5. False, end-to-end overlap
6. False, same number of Mos as atomic orbitals.
7. Bent, electron clouds form tetrahedral so angle ~ 109^0, but will be less because two lone pairs take up more space.
8. Five equivalent orbitals, sp^3d
9. Carbon atoms sp^2 hybridized. Six half-filled 2p orbitals combine to produce 6 Mos. Three bonding, which are filled and three antibonding which are not.

Self Test

1. d
2. c.
3. a.
4. d.

5. $:\ddot{O}-\ddot{O}=\ddot{O}:$ (and resonance structure) Unshared pair of electrons would give trigonal planar orbital geometry (120^0 bond angle) and bent molecular geometry with angle of slightly less than 120^0.

6. a. tetrahedral b. linear c. bent

7. Can form expanded octet with unfilled d orbitals spd hybridization.

8. BO = zero, unstable will have σ_{2s} bonding and antibonding orbitals filled.

9. Carbon will be sp^2 hybridized with p orbitals for pi bond.

10. N has five valence electrons. Will fill the σ_{2s} bonding and antibonding orbitals (4 electrons), then will fill two π_{2p} orbitals (4 electrons) and remaining two electrons will fill the σ_{2p} bonding orbital. BO = 3, diamagnetic, no unpaired electrons.

11. Electron clouds

12. Sigma bond

13. Molecular orbitals

14. Antibonding orbitals

15. Bond order

Chapter 11

Quiz A

1. CH_3CN
2. Water is polar and has H bonds.
3. (b) rate of condensation decreases
4. (b) solid to liquid
5. 2 atoms
6. A network covalent solid has covalent bonds throughout the solid.
7. (a) NH_3
8. 1-hexanol will have a higher boiling point because it will have hydrogen bonding.
9. Sublimation is the process of a solid going to a gas without first passing through a liquid phase.
10. 119 kJ

Quiz B

1. H_2O
2. NaH
3. (c) normal boiling point
4. 1 atom
5. (e) all of these
6. 47.2 kJ/mol

7. $MgCl_2$

8. Supercooling is when the temperature of a liquid drops below the freezing point without a solid appearing.

9. Polarizability is a measure of the ease with which electron charge density is distorted by an external field.

Quiz C

1. False, dipole-dipole and H bonding are stronger
2. True
3. False, larger atoms with more easily distorted electron clouds
4. False, H must be on O,N, or F.
5. True
6. c.
7. b.
8. b.
9. a.
10. c.

Self Test

1. f
2. h
3. g
4. j
5. e
6. b
7. c
8. a
9. d
10. i
11. HCN, stronger dipole-dipole forces
12. Stronger forces would have higher viscosity
13. HBr dipole-dipole forces primarily

    H_2O dipole-dipole forces and strong H bonding forces
14. 2.27 kJ/g x 18.0 g/mol = 40.9 kJ/mol
15. $\lambda = 2d\sin\theta$

    $204\text{pm} = 2d\sin(22.2^0)$

    $d = 270$ pm
16. K(s) → K(l) 2.4 kJ

$K(l) \rightarrow K(g)$ 82.5 kJ

$K(s) \rightarrow K(g)$ 84.9 kJ

Chapter 12

Quiz A

1. (a) The rate of crystallization is equal to the rate of dissolving.
2. Any two of the following: freezing point depression, osmotic pressure, boiling point elevation, vapor pressure lowering
3. 18 g KCl + 82 g H_2O
4. (b) 1.0 M $MgCl_2$
5. 1.29 m
6. 8.11 mmHg benzene, 67.9 mmHg toluene
7. 11.5 g KCl + 238.5 g H_2O
8. Heterogeneous. The intermolecular forces are different in the two liquids. Also, benzene is nonpolar and water is polar.
9. (b) doubles the solubility

Quiz B

1. 4.82 m
2. b) 0.01 M $CaCl_2$
3. 17.4 mmHg
4. -0.56 °C
5. Osmotic pressure is the pressure required to stop osmosis.
6. 0.139 m
7. Yes. H bond interactions would make it very soluble.
8. A supersaturated solution is one in which more solute is dissolved than is normally soluble in the solvent.
9. (b) become less soluble

Quiz C

1. False, osmosis not osmotic pressure
2. True
3. True
4. False, like dissolves like
5. False, solubility and pressure are directly related
6. False, liquid is the solvent

7. Assume 100 g of solution, 12.5g NaOH

   12.5 g x 1mol/40g = 0.3125 mol

   L of solvent = (100g – 12.5g) x 1mL/g = 87.5 mL

   0.3125mol/0.875 L = 3.57 m

8. Intermolecular forces are mainly H-bonding but since these are relatively strong intermolecular forces and ethanol is relatively small molecule a solution will form.

9. $\Pi = MRT$

   M = 12.0g x 1mol/180g x 1/1L = 0.0667 M

   Π = (0.0667 M)(0.0821 L atm $mol^{-1}K^{-1}$) x 293.15K = 1.61 atm

Self Test

1. h
2. f.
3. d
4. a
5. g
6. b
7. c
8. e
9. Assume 100g of solution, 36.5g HCl, 63.5g H_2O

   Mol HCl = 36.5 x 1mol/36.5g = 1.00 mol

   Kg H_2O = 63.5g x 1kg/10^3 g = 0.0635 Kg

   m = 1.00 mol solute/0.0635 Kg solvent = 15.7 m

   Molarity, 1.00 mol HCl from above calculation. To find volume of 100g of solution, use density. V = 100g x 1mL/1.18g x 10^{-3}L/mL = 0.0847 L

   M = mol/L = 1.00mol/0.0847 L = 11.8M

10. $\Delta T = K_b m$, m= 2.3 g x 1mol/58.4g x 1L/Kg = 0.0394 m, but for NaCl have two particles/mol so use 2 x m = 0.0788

    ΔT = (0.52)x(0.0788) = 0.041

    T = 100.041 ^{0}C

11. 1.3 mol x 58g/mol = 75.4 g NaCl, mass of solution = 1 Kg + 75.4 g = 1075.4 g

    % mass = 75.4 g/1075.4 g x 100 = 7.0 % NaCl by mass

12. $\Delta T = -K_f m$, m = -0.150/1.86 = 0.0806

    moles of solute = m x Kg solvent = (0.0806) (0.200 Kg) = 0.0161 moles

    15.0g = 0.0161 moles and MW = 15.0g/0.0161mol = 932 g/mol

13. $\Pi = MRT$

    Π= 2.5g/0.200L x 1mol/95.2g x 0.0821 L atm K^{-1} mol^{-1} x 298.15 x 3 particles/mol

    $MgCl_2$ dissociates to 3 particles so use factor of 3.

    Π= 9.64 atm = 9.6 atm

14. $\Delta P = X_2 P_i^0$, where X_2 = mol fraction of solute

    g NaOH x 1mol/40.0g = 0.030 mol

    100.0 mL H^2O x 1g/mL x 1 mol/18.0g = 5.56 mol

    X_2 = 0.030 mol/(0.030 mol + 5.56mol) = 0.00537, for NaOH, 2 particles, therefore X = 0.0107

    ΔP = (0.0107)(42.175 mmHg) = 0.451 mmHg

    P = 42.175 – 0.451 mm Hg = 41.724 mmHg

15. colloids
16. solvent-solvent, solute-solvent, solute-solute
17. less
18. an increase

Chapter 13

Quiz A

1. x = 2, y = 1
2. rate = $k[A]^2$
3. 0.231 s^{-1}
4. (c) 20 min
5. (a) the value of ΔH
6. (d) rate constant remains unchanged
7. third order
8. An activated complex is a transitory species formed through collisions between molecules having sufficient kinetic energies and proper orientation.
9. second order
10. 10 min.

Quiz B

1. (b) doubling B
2. 0.46 s
3. (c) is not changed by the reaction
4. (b) the reaction is a first-order process
5. The kinetic energy of the molecules increases and the number of collisions with sufficient energy increases.

6. A reaction mechanism consists of a series of elementary steps which yield the observed net equation and the observed rate law.
7. 0.14 s^{-1}
8. $M^{-2}s^{-1}$
9. 0.058 s^{-1}
10. second order

Quiz C

1. False, a single step in a reaction mechanism
2. True
3. True
4. False, exponents of the concentrations in the rate law
5. False, determine experimentally
6. False, depends on order of reaction as to how it varies
7. True
8. True
9. True
10. False, maximum

Self Test

1. second order
2. rate constant and initial concentration
3. increases
4. activation energy
5. transition state
6. c
7. c
8. d
9. b
10. d
11. Step 1
12. C is an intermediate.
13. Rate = k[A][B]
14. No. If a single elementary step, exponent of [A] would be 2.
15. $\ln (k_2/k_1) = (E_a/R)(1/T_1 - 1/T_2)$
    $\ln (k_2/k_1) = 86.6$, $(k_2/k_1) = 4.07 \times 10^{37}$

Chapter 14

Quiz A

1. $K_c = 1/[Cl_2]^3$
2. product decreases; reactants increase
3. favors reactants
4. no effect
5. (d) no change
6. A reaction is said to be at equilibrium when the forward and reverse rates of reaction are equal and there is no net change in the concentrations of reactants or products.
7. (d) the square
8. $K_p = K_c(RT)^{\Delta n}$gas
9. (a) the reaction goes to completion
10. A heterogeneous equilibrium is one which involves reactants or products in different phases, such as gases and solids.

Quiz B

1. $K_c = [NH_3]^2/[N_2][H_2]^3$
2. $K_p = K_c(RT)^{-2}$
3. reactants increase, products decrease
4. (b) forward reaction doesn't occur to any significant extent.
5. to the left toward reactants.
6. no effect on equilibrium
7. Decreasing the volume will favor products because there are more moles of gas on the reactant side of the equilibrium.
8. Increasing temperature will shift the equilibrium toward products (to the right).
9. 27
10. A homogeneous equilibrium is one in which reactants and products are in the same physical state.

Quiz C

1. True
2. False, catalyst speeds up reaction
3. True
4. False, shift to right
5. True
6. False, two phases are involved

7. False, decrease
8. True

Self Test

1. changes
2. the reciprocal
3. not affect
4. reaction quotient
5. Le Chatelier's principle
6. Decreases
7. $K = [SO_2]/[O_2]$
8. Shift to left
9. No change
10. No change
11. Shift to right
12. $K' = [O_2]/[SO_2]$
13. $K_p = K_c(RT)^{\Delta n} = (0.24)[(0.0821)(573)]^1 = 11.29$
14. $Q = [0.5][2.3]/[1.2] = 0.96$ $Q>K_c$ not at equilibrium
15. $PCl_5 \leftrightarrow Cl_2 + PCl_3$

    At equilibrium

    $1.2 + x$ $\quad 2.3 - x \quad 0.5- x$

    $0.24 = (2.3\text{-}x)(0.5\text{-}x)/(1.2\text{+}x)$

    $0.288 + 0.24x = 1.15\text{-}2.8x + x^2$

    rearranging: $x^2 - 3.04x + 0.862 = 0$

    Using the quadratic equation, $x = 0.316$
16. Equilibrium concentrations: $[PCl_5] = 1.52, [Cl_2] = 2.0, [PCl_3] = 0.2$

Chapter 15

Quiz A

1. (b) acid
2. HBr
3. neutral
4. CN^- base/HCN conjugate acid; H_2O acid/OH^- conjugate base
5. (b) pH = 10.0
6. $NH_3 + H_2O \leftrightarrow NH_4^+ + OH^-$

    base (1) acid (2) acid(1) base (2)

7. acidic
8. acetic acid
9. The self ionization of water will contribute 10^{-7} $[H_3O^+]$ ions. The total $[H_3O^+]$ will be slightly higher than this and therefore the pH will be < 7.
10. A substance that is added to a titration solution in small amounts and changes color at the equivalence point.
11. pH = 9.08

Quiz B

1. pH = 1.70
2. proton acceptor
3. basic
4. less than 7
5. pOH = 9.0
6. pH = 12.3
7. The equivalent point is the condition where the reactants are in stoichiometric proportions.
8. smaller pKa value
9. neutral
10. $NH_3 + H_2PO_4^- \text{ <--> } NH_4^+ + HPO_4^{2-}$

    base (1) acid (2) acid (1) base (2)

Quiz C

1. True
2. False, electron pair acceptor
3. True
4. False, 14
5. True
6. False, a weak base is weakly ionized
7. True
8. True
9. True

Self Test

1. d
2. b
3. f
4. h

5. g
6. a
7. e
8. c
9. $1.5 \times 10^{-5} = [CH_3(CH)_2COO^-][H^+]/[CH_3(CH)_2COOH] = x^2/(0.15-x)$
   Assume x is small, then $x = 1.5 \times 10^{-3}$ and taking the –log, pH = 2.8
10. pH = 10.2
    pOH = 14 – 10.2 = 3.8 and taking the antilog, $[OH] = 1.58 \times 10^{-4}$
11. Solve for $[OH^-]$. $[OH^-] = 2.1 \times 10^{-5}$, pH = 14-pOH = 9.3
12. pH = 5.6 = 14 – pOH
    pOH = 8.4 = $-\log[OH^-]$
    $[OH^-] = 4 \times 10^{-9}$
13. H_3PO_4, first ionization is easier and therefore a stronger acid.
14. HCl
15. $0.01M = [OH^-]$
    $[H^+] = K_w/[0.01M] = 1. X 10^{-14}/0.01 = 1.0 \times 10^{-12}$

Chapter 16

Quiz A

1. $K_{sp} = [Ca^{2+}]^3[PO_4^{3-}]^2$
2. 8.1×10^{-11}
3. (b) add HCl
4. (b) decrease $[OH^-]$
5. (b) The concentration of a substance is 0.1% of the initial concentration.
6. (d) is just saturated
7. $K_{sp} = [Pb^{2+}][F^-]^2$
8. 2.61×10^{-3} M
9. $Pb_3(AsO_4)_2 \longleftrightarrow 3\, Pb^{2+} + 2\, AsO_4^{3-}$
10. 9.5×10^{-6} M

Quiz B

1. 1.4×10^{-5}
2. $K_{sp} = [Bi^{3+}]^2[S^{2-}]^3$
3. (a) more solid dissolves
4. (b) 0.100 M HCl
5. 1.4×10^{-17}

6. CrF_3 <--> $Cr^{3+} + 3\ F^-$
7. 7.1×10^{-7} M
8. A complex ion is a polyatomic cation or anion composed of a central ion bonded to other molecules or ions.
9. A precipitate forms
10. The formation constant of a complex ion is the equilibrium constant describing the formation of the complex ion from a central atom and its ligands.

Quiz C

1. Solubility product constant
2. Equal
3. Decrease
4. $[Ca^{2+}][F^-]^2$
5. amphoteric oxides
6. precipitation
7. formation constant
8. qualitative analysis
9. increases dramatically
10. Group I ($AgCl$, Hg_2Cl_2, $PbCl_2$)

Self Test

1. $K_{sp} = [Pb^{2+}][Cl^-]^2 = 1.6 \times 10^{-5} = (s)(2s + 0.1)^2$ Assume 0.1 >> s, $(s)(0.1)^2 = 1.6 \times 10^{-5}$

   $s = 1.6 \times 10^{-3}$
2. $K_{sp} = (s)(3s)^3 = 27s^4 = 1.3 \times 10^{-33}$ $s = 2.6 \times 10^{-9}$
3. HCl will react with OH^- and by LeChatelier's principle, the solubility will be increased. (Product is removed).
4. $K_{sp} = (Mg)(OH^-)^2 = (s)(2s)^2 = (1.12)(2.24)^2 \times 10^{-12} = 5.62 \times 10^{-12}$
5. pH 8.0 because OH^- concentration would be lower and OH^- is a common ion.
6. $K_f = [Ag(CN)_2^-]/[Ag^+][CN^-]^2$ rearranging, $[Ag^+]/[Ag(CN)_2^-] = 1/K_f[CN^-]^2$ $= 1/(1.0 \times 10^{21})(0.02)^2 = 2.5 \times 10^{-18}$
7. At pH 8.0, pOH = 14 – 8.0 = 6.0 = $-\log[OH^-]$ $[OH^-] = 1.0 \times 10^{-6}$

   $K_{sp} = (s)(3s + 1.0 \times 10^{-6})^3$ approximate as $(s)(1.0 \times 10^{-6})^3 = 1.3 \times 10^{-33}$

   $s = 1.3 \times 10^{-15}$
8. Larger K_f is more stable
9. d) water
10. Concentrations after mixing are: 0.5×10^{-4} M $FeCl_3$ pOH = 14 – 9 = 5.0 $[OH^-] = 1.0 \times 10^{-5}$ and this is diluted by two so concentration is 1/2, or 0.5×10^{-5}

$Q = (0.5 \times 10^{-4})(5 \times 10^{-6})^3 = 6.25 \times 10^{-21}$

$Q > K_{sp}$ Therefore a precipitate will form.

Chapter 17

Quiz A

1. negative
2. (d) spontaneous
3. (c) free energy
4. (a) low temperature
5. -2.89 kJ/mol
6. The free energy is a thermodynamic term that describes the spontaneity of a reaction.
7. The entropy of a perfect crystal at 0 K is zero.
8. The entropy is positive.
9. 1630 K
10. A nonspontaneous reaction or change will not occur unless some external action is applied.

Quiz B

1. positive
2. $\Delta H < 0$
3. (d) nonspontaneous
4. $K_{eq} = \frac{P^2_{HBr}}{P_{H_2} P_{Br_2}}$
5. $K_{eq} = 5.4 \times 10^4$
6. A spontaneous change occurs in a system that is left to itself. No action from outside the system is necessary for the change to occur.
7. The enthalpy of the reaction is negative.
8. Entropy is a measure of the disorder of a system.
9. + 42.2 kJ
10. (c) zero

Quiz C

1. False. Spontaneous reactions can be exothermic or endothermic.
2. True
3. False. Entropy increases
4. True
5. False. Systems spontaneously become more disordered.

6. True
7. False. The Rate of reaction refers to the kinetics, not whether the reaction is spontaneous.
8. True
9. False. Both the enthalpy and the entropy account for the direction of the reaction.

Self Test

1. c
2. e
3. d
4. b
5. f
6. a
7. ΔS will be negative, therefore $-T\Delta S$ will be a positive term, but ΔH is negative so $|\Delta H| > |\Delta S|$ and $\Delta G < 0$.
8. $T = \Delta H/\Delta S$

   $\Delta H = \Delta H_f[CaO(s)] + \Delta H_f[H_2O(g)] - \Delta H_f[Ca(OH)_2(s)] = 109.3$ kJ

   ΔS found in the same way using S_f. $\Delta S = +145.1$ J/K

   $T = 1.093 \times 10^5$ J/145.1 JK^{-1} = 753 K
9. $\Delta G = \Delta H - T\Delta S = -RT\ln K$

   First solve for ΔG

   $\Delta G = 1.093 \times 10^5$ J – (500K)(145.1 JK^{-1}) = 3.68×10^4 J = $-RT\ln K$ solve for K

   $\ln K = 3.68 \times 10^4$ J/(-8.31)(500K) = -8.83

   $K = 1.43 \times 10^{-4}$
10. a, c, d

Chapter 18

Quiz A

1. +0.771 V
2. (c) oxidation occurs
3. (c) spontaneous in opposite direction
4. $Zn(s) \text{ --> } Zn^{2+}(aq) + 2\,e^-$

   $Cu^{2+}(aq) + 2\,e^- \text{ --> } Cu(s)$
5. (b) 1 C/s
6. $Pt|Fe^{2+}(aq), Fe^{3+}(aq) \,||\, Cr_2O_7^{2-}(aq), Cr^{3+}(aq)\,|Pt$

7. 0.00 Volts
8. No. The reduction potential will be greater for Cl_2.
9. $2 Ag(s) + Cl_2 \text{-->} 2 Cl^- + 2 Ag^+$ $E^o_{cell} = +0.56$ V
10. Corrosion of a metal is its oxidation in the presence of oxygen.

Quiz B

1. (b) reduction occurs
2. Fe(s) anode, Cu(s) cathode
3. $Fe(s) \text{-->} Fe^{2+}(aq) + 2 e^-$

   $Cu^{2+}(aq) + 2 e^- \text{-->} Cu(s)$
4. (d) Li will displace $H_2(g)$ from H_2O and no Na will form.
5. (b) $E_{cell} > E^o_{cell}$
6. A galvanic or voltaic cell is an electrochemical cell that contains an oxidation-reduction reaction.. The reaction is spontaneous and produces an electrical current.
7. (b) Faraday's constant
8. $\Delta G = -n \times F \times E_{cell}$
9. A battery is a number of electrochemical cells wired together.
10. The iron object to be protected is connected to an active metal. Oxidation will occur at the active metal first and the metal will dissolve. The iron surface will acquire electrons from the active metal and the reduction half-reaction will be favored.

Quiz C

1. False, Nonspontaneous reaction is made to occur by applying an electric current
2. True
3. True
4. False. Must measure the potential against a reference electrode
5. True
6. False. Corrosion is the oxidation of metal
7. True
8. True
9. True

Self Test

1. cathode
2. corrosion
3. in their standard states

4. reversed
5. below it
6. temperature and composition
7. Nernst
8. Cathode
9. No. half reaction $2I^- \rightarrow I_2 + 2e^-$

   $SO_4^{2-} + 4H^+ + 2e^- \rightarrow H_2SO_4 + H_2O$

   I^- lies above SO_4^-, the oxidizing agent, therefore I^- cannot reduce SO_4^{2-} and the reaction is not spontaneous.
10. $4Fe^{2+} \rightarrow 4Fe^{3+} + 4e^-$ $E = -0.77$ V

    $O_2 + 4H^+ + 4e^- \rightarrow 2H_2O$ $E = +1.23$ V

    $E_{cell} = 0.46$ V $\Delta G = -nFE_{cell} = (-4)(96500)(0.46) = -180$ kJ

    $\text{Log } K = nE_{cell}/0.0592 = (4)(0.46)/0.0592 = 31$ $K = 10^{31}$
11. $2Cl^- \rightarrow Cl_2 + 2e^-$

    5.0 A x 25 min x 1 Cs^{-1}/A x 60 s/min = 7500 C

    7.5×10^3 C x 1 F/96,500 C x 1 mol Cl_2/2F = 0.039 mol Cl_2

    0.039 mol Cl_2 x 70.0 g/mol = 2.76 g Cl_2
12. Dry cell, mercury, nickel-cadmium, fuel cells. Batteries can be recharged if solid products of the electrode reaction adhere to the surface of the electrodes.
13. $F_2 < H_2 < Ni < Zn$
14. Cell emf will increase, due to decrease in $[H^+]$

Chapter 19

Quiz A

1. False. Mass number is the same. There is a small difference in the exact mass.
2. True
3. False. Positron has a mass of an electron.
4. True
5. False. Even numbers are more stable.
6. False. Number of protons and neutrons that confer extra stability.
7. False. See text for examples.
8. True
9. True
10. False. Many different nuclei are formed.

Quiz B

1. c
2. d

3. b
4. b
5. a
6. d
7. a

Self Test

1. e
2. d
3. g
4. a
5. f
6. c
7. b
8. a. $^{104}_{46}Pd$
   b. $^{0}_{1}e$
9. a $\rightarrow$ $^{1}_{0}e + ^{74}_{34}Se$
   b $\rightarrow$ $^{0}_{-1}e + ^{73}_{32}Ge$
10. ln (4.0 dpm/g)/(15.3 dpm/g) = -0.693 (t/5715 y)
    t = 11,000 y

Chapter 20

Quiz A

1. a. sodium carbonate b. calcium oxide c. lithium hydride
2. $CaO + 2\ HCl \text{-->} CaCl_2 + H_2O$
3. 2 MgO
4. See Table 20.7
5. electrolysis of water, water gas reactions, and the reforming of hydrocarbons.
6. Ca^{2+}, Mg^{2+}, Fe^{2+}, HCO_3^- and SO_4^{2-}
7. Group 2A have smaller atomic radii and greater ionization energies.
8. a. Calcium carbonate b. Barium hydroxide

Quiz B

1. $BaO + H_2O \text{-->} Ba(OH)_2$
2. $CaBr_2{\cdot}6H_2O$

3. $CaCO_3(s)$ --> $CaO(s) + CO_2(g)$ This reaction is important because it leads to the production of pure $CaCO_3$, an important industrial material used as a filler among other uses.
4. A group 1A metal is larger than its corresponding group IIA metal.
5. Elements on the diagonal from each other sometimes have similar properties, due to the charge density. This most often occurs with the first member of a group. Li and Mg are an example of elements that have a diagonal relationship.
6. $Zn(s) + 2H^+(aq)$ --> $Zn^{2+}(aq) + H_2(g)$
7. $3H_2(g) + N_2(g)$ --> $2NH_3(g)$
8. In a hydrogenation reaction, hydrogen is added to multiple bonds in another compound.
9. No reaction occurs.
10. CaO, calcium oxide

Quiz C

1. True
2. True
3. False. Magnesium does not react with cold water
4. False mostly calcium carbonate
5. True
6. True
7. True
8. False. Ion exchange is usually used.

Self Test

1. $M + X_2 \rightarrow MX_2$
2. The Dow process is the main process used. $Ca(OH)_2$ reacts with Mg^{2+} to produce $Mg(OH)_2$ which reacts with HCL producing $MgCl_2$. $MgCl_2$ is converted to Mg(s) by electrolysis.
3. $2Al(s) + 6HCl \rightarrow 2AlCl_3 + 3H_2(g)$
4. $2Li + 2H_2O \rightarrow 2Li(OH) + H_2$
5. Calcium carbonate is used as building material and as a chemical raw material.
6. Ca^{2+}, Mg^{2+}
7. $CaCO_3 \rightarrow CaO + CO_2$ (heat)
8. $CH_4(g) + H_2O \rightarrow CO(g) + 3H_2(g)$ (catalyst)
9. Used in alloys, flash bulbs, batteries
10. $C(s) + H_2O(g) \rightarrow CO(g) + H_2(g)$

Chapter 21

Quiz A

1. (d) H bonding
2. H_2O forms H bonds.
3. $Na_2B_4O_7 \cdot 10H_2O$
4. NH_2NH_2
5. 1. Bleach 2. preparation of chlorinated organic compounds
   3. Preparation of chlorinated inorganic compounds.
6. $2\ Al(s) + 6\ H^+(aq) \rightarrow 2\ Al^{3+}(aq) + 3\ H_2(g)$
7. $[Ar]3d^{10}4s^2$ for Ga^+ and $[Ar]3d^{10}$ for Ga^{3+}
8. Acetylene
9. Lead storage batteries
10. Allotropes are two or more different forms of the same element. Some examples are graphite and diamond and red and white phosphorus.

Quiz B

1. Al
2. Se
3. Si
4. $NH_2NH_2 + 2\ H_2O \rightarrow NH_3NH_3^{2+} + 2\ OH^-$
5. (b) xenon
6. F_2
7. $MnO_2(s) + 4\ HCl(aq) \rightarrow MnCl_2(aq) + 2\ H_2O + Cl_2(g)$
8. Boron
10. Oxygen is the most nonmetallic.

Quiz C

1. borax and kernite
2. sulfuric acid
3. boric acid
4. not reactive
5. carbide
6. ammonia
7. F used in toothpaste, plastics; Cl in bleaching, disinfectant; Br photographic film
8. Si forms significantly fewer hydrides. Si hydrides are more reactive. Silanes behave more like metals than alkanes.
9. A compound of two or more halogens.

10. ns^2 electrons in the valence shell of Group 3A, 4A and 5A elements. These electrons remain in the valence shell after loss of np electrons, e.g. Sn^{2+}, Pb^{2+}, Bi^{3+}

Self Test

1. The valence shell is the p shell.
2. Kr, Xe, Rn
3. $2BCl_3 + 3H_2 \rightarrow 2B + 6HCl$
4. $Al_2O_3(s) + 2OH^- + 3H_2O \rightarrow 2[Al(OH)_4]^-$
5. ns^2np^3 Electronegativity decreases, more metallic lower in the group.
6. $CuF_2 + H_2SO_4 \rightarrow 2HF + CaSO_4$
   $2HF \rightarrow H_2 + F_2$
   $2NaCl + 2H_2O \rightarrow 2NaOH + H_2 + Cl_2$
7. Ammonia is reacted with oxygen to give NO. Additional oxygen is introduced to give NO_2. NO_2 is dissolved in water to give nitric acid.
8. $3H_2 + N_2 \rightarrow 2NH_3$ Fertilizer is the main use.
9. HF is a weak acid because of strong H bonding.
10.

+5	NO_3^-
+4	NO_2
+3	NO_2^-
+2	NO
+1	N_2O
0	N_2
-1	NH_2Cl
-2	N_2H_4
-3	NH_3

Chapter 22

Quiz A

1. (d) all of these
2. a half-filled d subshell is very stable
3. $Cr_2O_7^{2-}$
4. +6
5. Iron coated with zinc.
6. tetraamminedichlorocobalt(III) ion
7. +3
8. $[FeCl_6]^{3-}$

Quiz B

1. Yes, 2 unpaired electrons
2. Yes. Cis and trans isomers are possible.
3. Optical isomers are molecules and ions that differ in their optical activity. They have nonsuperimposable mirror images.
4. (c) Ti
5. $[Ar]3d^5$
6. $Mn(s) + 2\ HCl(aq) \dashrightarrow MnCl_2(aq) + H_2(g)$
7. $Cr_2O_3 + 2\ Al \dashrightarrow Al_2O_3 + 2\ Cr$
8. Na_2CrO_4
9. +5
10. Au

Quiz C

1. (b) the transition elements
2. Fe, Co and Ni
3. Iron
4. These elements have filled d-subshells.
5. $[Ar]3d^2$
6. Two applications of coordination chemistry are sequestering of metal ions and qualitative analysis.
7. Many complex ions are colored because the energy differences between d orbitals match the energies of components of visible light.
8. The oxidation state of Ni is 0 and the coordination number is 4.
9. A coordinate covalent bond is a bond formed by sharing a pair of electrons between two atoms. The electron pair is provided by only one member of the covalent bond.

Self Test

1. (a) Fe, Co and Ni
2. CrO_3
3. (a) filling the 4f subshell
4. six
5. bis(ethylenediamine)copper(II) ion
6. No. They are the same ion written two different ways.
7. +2
8. tetraamminedichlorocobalt(III) ion.

9. Structural isomers are isomers which differ in the ligands that are attached to the central atom and/or the donor atoms through which they are bonded. Their formulas are the same.

Chapter 23

Quiz A

1. benzene
2. penta
3. CO_2H
4. Ortho
5. Meta
6. Amines and carboxylic acid
7. Adenine, guanine, thymine, cytosine
8. Base, sugar, phosphate
9. Infra-red
10. Odd

Quiz B

1. False, they are defined by properties
2. False, glycerol and fatty acid
3. False, no double bonds.
4. False, triple bond and suffix is yne
5. True
6. True
7. False. Act as bases
8. False. Generally weak acids.

Quiz C

1. CH_2BrCH_2Br
2. CH_3CH_2OH
3. CH_3CH_3
4. $CH_3CH_2CH_2I$
5. $CH_3CH{=}CH_2$
6. $CH_3CH_2CH_2OCH_2CH_2CH_3$
7. $CH_3COCH_2CH_3 \rightarrow CH_3COO^-Na^+ + CH_3CH_2OH$

   ||

   O

8.

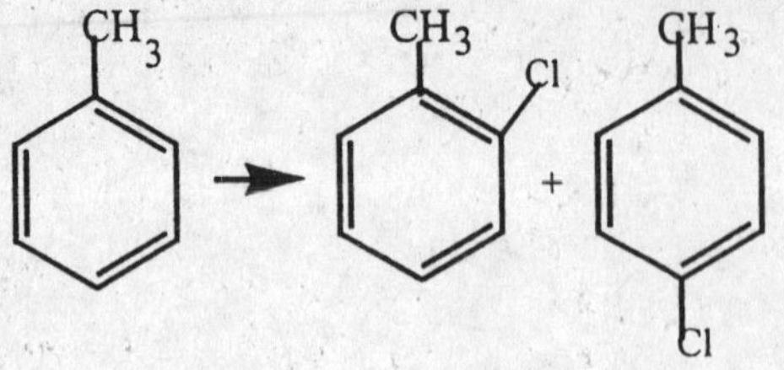

CH_3 CH_3 Cl + CH_3 Cl

9.

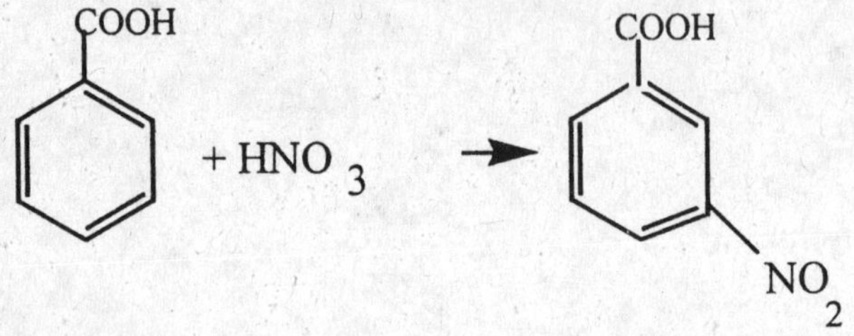

COOH + HNO_3 → COOH NO_2

Self Test

1.

(a) 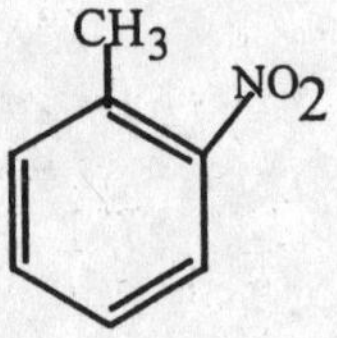 CH_3 NO_2

(b) 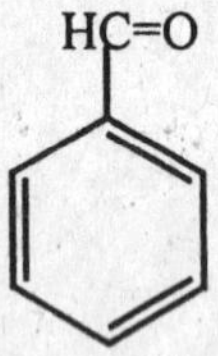 HC=O

(c) $H_3CCH_2CHCl\text{-}C{\equiv}CH$

(d) $CH_3CHCHCH_2OH$ (with CH_3 on C-2 and CH_3 on C-3)

$$\begin{array}{l} CH_3\underset{\displaystyle CH_3}{\underset{|}{C}}H\underset{\displaystyle CH_3}{\underset{|}{C}}HCH_2OH \end{array}$$

(e) $CH_3CH_2COCH_2CH_3$

$$CH_3CH_2\overset{}{\underset{\displaystyle O}{\underset{\|}{C}}}CH_2CH_3$$

2. Amino acids are the basic units. $RC(NH_2)COOH$ peptide: $H_2NCHCNHCCOOH$

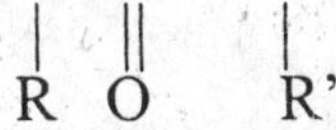

3. IR and nmr
4. Ala CH_3-- ; Ile $CH_3CH_2CH(CH_3)$- ; Ser $HOCH_2$-
5. TAAGGC
6. Delocalized π electrons
7. Either the atomic number or the mass number must be odd.

Chapter 24

Quiz A

1. metallurgy
2. a sea of electrons
3. the same as
4. semiconductor
5. the band gap
6. decreases
7. plastics, elastomers, fibers
8. addition, condensation
9. biopolymers

Quiz B

1. False Describes the electrical conductivity of metals.
2. True
3. False. Doped material has an added acceptor or donor substance.
4. False Large difference of energy between bands.
5. False Band gap is the difference between conduction and valence bands.
6. False Addition polymerization includes all of the monomer.
7. True
8. False. Has a high tensile strength.

Quiz C

1. cassiterite, SnO_2
2. impure iron formed in a blast furnace
3. ZnS and some $ZnCO_3$
4. $2PbS + 3O_2 \rightarrow 2PbO(s) + 2SO_2(g)$

   $PbO(s) + C(s) \rightarrow Pb(l) + CO(g)$

   $PbO(s) + CO(g) \rightarrow Pb(l) + CO_2(g)$

5. oxides, $ZnCO_3 \rightarrow ZnO + CO_2$
6. Electrons must be free to move from one orbital to another in the band. If the band is completely filled electrons have no place to go.
7. A polymer is a macromolecular material built up of monomer units. Examples: proteins, cellulose, plastics, nucleic acids, etc.

Self Test

1. Yields a pure metal and chemical methods don't work well because of highly reactive nature of the metals.
2. Crystal made up of n atoms with overlapping atomic orbitals will have 1/2 n bonding and 1/2 n antibonding MOs. Alkali metals have n valence electrons, n orbitals so half filled. Because close in energy electrons move.
3. The 3p band lies at lower energy than 3s. Some of 3s electrons found in 3p band and therefore 3s and 3p are only partially filled.
4. $n(H_2C{=}CH_2) \rightarrow -CH_2CH_2CH_2CH_2-$ (unit repeated n times, shown are two units)
5. $RC(NH_2)COOH + R'C(NH_2)COOH \rightarrow H_2NCH\text{-}C\text{-}NH\text{-}CH\text{-}COOH$

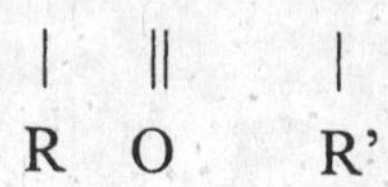

6. n-type involves movement of electrons in the conduction band, p-type involves migration of positive holes.
7. High density—great rigidity, strength, high melting point, consists mainly of linear polymers.

   Low density—semirigid, low melting point, consists mainly of branched molecules
8. cellophane is made up of cellulose xanthate which is forced through a narrow slit.
9. $-CH_2\text{-}CCl_2\text{-}CH_2\text{-}CCl_2-$
10. Propagation ends when a molecule is produced that no longer has an unpaired electron, possibly by combination of two radicals through an electron pair bond.

Chapter 25

Quiz A

1. False Nitrogen is the most abundant
2. True
3. False Weather occurs in the troposphere.
4. False. Refers to trapping of radiant energy by CO_2 and other gases.
5. True
6. True
7. True
8. False. Higher concentration of ozone.
9. False Fluorine prevents decay.

10. True

Quiz B

1. Biochemical oxygen demand
2. Microorganisms
3. Chlorine
4. Sulfur oxides, nitrogen oxides
5. Absorbing UV radiation
6. Chlorofluorocarbons
7. Toxicology
8. Inactivate enzymes
9. Oxygen
10. Partial pressure H_2O vapor/ vapor P of H_2O

Self Test

1. d
2. e
3. c
4. f
5. h
6. a
7. b
8. g
9. Troposphere—nearest Earth, weather occurs; next is stratosphere—contains ozone layer; ionosphere—molecules absorb energy and dissociate into ions
10. CO_2 and other gases absorb and trap energy radiated by the Earth's surface. Increase in Earth's temperature could lead to drier agricultural areas, higher sea levels from ice melting, changes in forestation and plant and animal life.
11. 1) addition of slaked lime and flocculating agent to remove dirt and bacteria
    2) filtering 3) aeration removes odorous compounds 4) chlorination to kill any remaining bacteria.
12. nitrogen oxides, hydrocarbons
13. CO interferes with oxygen transport; heavy metals inactivate enzymes; CN shuts down cell respiration.